150387498 3

AF598495

New Directions in the Theory of Graphs

ACADEMIC PRESS RAPID MANUSCRIPT REPRODUCTION

Proceedings of the
Third Ann Arbor Conference on Graph Theory
Held at the University of Michigan
October 21-23, 1971

New Directions in the Theory of Graphs

Edited by

Frank Harary

Department of Mathematics
The University of Michigan
Ann Arbor, Michigan

ACADEMIC PRESS New York and London 1973

ACADEMIC PRESS, INC.
111 Fifth Avenue, New York, New York 10003

United Kingdom Edition published by
ACADEMIC PRESS, INC. (LONDON) LTD.
24/28 Oval Road, London NW1

LIBRARY OF CONGRESS CATALOG CARD NUMBER: 72-77368

PRINTED IN THE UNITED STATES OF AMERICA

TO Miriam
Natalie
Judith
Thomas
Joel
Ilene

CONTENTS

CONTRIBUTORS

Morton Brown
Department of Mathematics, The University of Michigan, Ann Arbor, Michigan 48104

William Brown
Department of Mathematics, McGill University, Montreal 110, Canada

Václav Chvátal
Mathematics Research Center, University of Montreal, Casa Postale 6128, Montreal, Canada

Robert Connelly
Department of Mathematics, Cornell University, Ithaca, New York 14850

Paul Erdös
Mathematical Institute of the Hungarian Academy of Sciences, Realtanoda u. 13-15, Budapest V, Hungary

Frank Harary
Department of Mathematics, The University of Michigan, Ann Arbor, Michigan 48104

Marshall Hestenes
Department of Mathematics, Michigan State University, East Lansing, Michigan 48823

Abbe Mowshowitz
Department of Computer Science, University of British Columbia, Vancouver, British Columbia, Canada

Crispin Nash-Williams
Department of Mathematics, University of Aberdeen, Aberdeen, Scotland

Edgar Palmer
Department of Mathematics, Michigan State University, East Lansing, Michigan 48823

Richard Rado
Department of Mathematics, University of Reading, Reading, England

Robert Robinson
Department of Mathematics, The University of Michigan, Ann Arbor, Michigan 48104

Allen Schwenk
Department of Mathematics, The University of Michigan, Ann Arbor, Michigan 48104

Vera Sós
Mathematical Institute of the Hungarian Academy of Sciences, Realtanoda u. 13-15, Budapest V, Hungary

William Tutte
Department of C & O, University of Waterloo, Waterloo, Ontario, Canada

PREFACE

This book contains the proceedings of the Third Ann Arbor Conference on Graph Theory. The first such conference was never recorded for posterity, a fact which has caused some confusion to dedicated librarians in various parts of the world. The Second Ann Arbor Conference on Graph Theory was published under the title "Proof Techniques in Graph Theory," and was also published by Academic Press (1969). Like the present volume, all of the royalties of the preceding book were assigned to SIAM, The Society for Industrial and Applied Mathematics, for the worthy purpose of establishing "The George Pólya Prize in Combinatorics and Its Applications." Additional volumes, also published by Academic Press, have had their royalties assigned to the Pólya Prize; these comprise:

W. T. Tutte (editor), "Recent Results in Combinatorics" (1969)
R. C. Read (editor), "Graph Theory and Computing" (1972)

The theme of this conference is the demonstration of the interaction between graph theory and other branches of mathematics. From this viewpoint, a brief description of the contents is in order. My chapter on the history of the theory of graphs describes what are in my opinion the twelve most important and classical theorems of this theory. The next chapter, by Morton Brown and Robert Connelly, shows the outcome of an investigation into a subtle and delicate problem in graph theory, the characterization of graphs with constant link, when attacked by mathematicians with a powerful background in topology. In the third chapter by William Brown, the venerable Paul Erdös, and Vera Sós, hypergraphs are studied for their extremal properties. (A celebrated anecdote attributed to the famous Hungarian mathematician, Paul Turán, the conjecture that there are no truly beautiful female mathematicians but that he married a counterexample and her maiden name was Vera Sós.)

Then Václav Chvátal develops a generalization of the concept of a hamiltonian graph in the network setting which has proved useful in algorithms on combinatorial optimization. Algebraists have recently become quite interested in graph theory because they have found that certain rank 3 graphs help them to construct new finite simple groups; the method of construction is described by Marshall Hestenes. Abbe Mowshowitz explores some unexpected interaction between graphs and matrices. Then Crispin Nash-Williams, in an article which could itself be expanded into a separate monograph, explores unexplored and semiexplored territories in graph theory.

In the next contribution, Edgar Palmer discusses graphical enumeration methods and in the process derives another new result, the number of digraphs which are both self-converse and self-complementary. (He and I are now in the midst of proofreading our first joint book, "Graphical Enumeration," also to be published by Academic Press.) Richard Rado develops several results involving isomorphisms between hypergraphs. Robert Robinson, the hero of graphical enumeration, counts labeled digraphs with given strong components, which includes acyclic digraphs and strong digraphs as special cases. Using conventional enumeration techniques for trees in an unconventional manner, Allen Schwenk demonstrates the asymptotic result that almost all trees are cospectral. Alphabetically last but not least, William Tutte investigates an axiomatic formalization of a map in the plane.

Several expressions of gratitude are in order. All the speakers are to be thanked for the high level of their lectures and the resulting contributions. The participants displayed keen interest at the conference itself. The staff of Academic Press has been kind in their support of all phases of the production of this book. The Applied Mathematics Directorate of the Air Force Office of Scientific Research provided partial financial support. My unique doctoral student, Allen J. Schwenk, was most helpful with his insightful comments and proofreading. Finally, Joyce Lemen was terrific for the speed, accuracy, patience, and good humor with which she typed all twelve chapters on the special paper provided by the publishers.

Ann Arbor, Michigan Frank Harary

New Directions in the Theory of Graphs

ON THE HISTORY OF THE THEORY OF GRAPHS

Frank Harary

My own view of the history of a branch of mathematics is one which presents its most important theorems. To achieve this for graph theory, I would choose what I consider to be the twelve most "classical" theorems. Let us begin with the four outstanding pre-twentieth-century ones, namely

1)	Euler	(1736)
2)	Kirchhoff	(1847)
3)	Cayley	(1857)
4)	Heawood	(1890)

1. Euler.

Since Euler's day, an extra bridge has been added to the ancient seven bridges of Königsberg, now Kaliningrad, see Euler [3,4]. If this spans the whole river without access to an island, then the original graph (with the vertices or points for land areas and the edges or lines representing bridges) and the new graph are

original graph

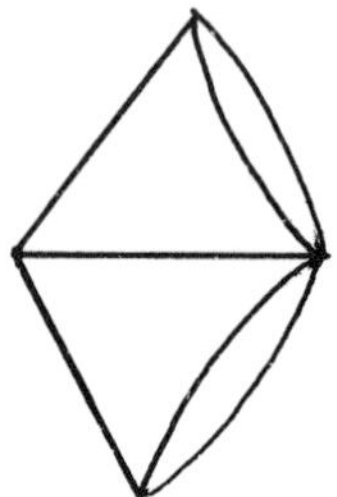

new graph

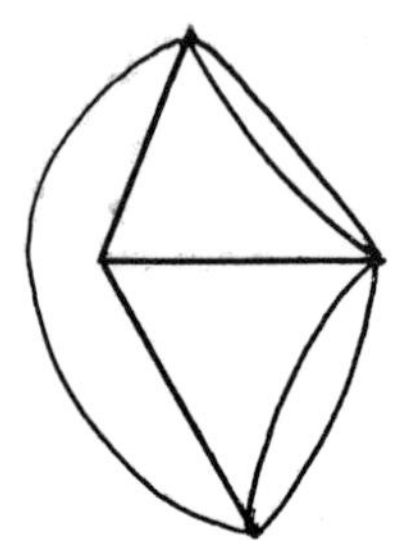

Figure 1

(Strictly speaking, both of these are multigraphs.) The new graph is connected, has just two points of odd degree, and therefore it can be shown to possess an open trail containing each of the lines just once; but the graph is still not "eulerian" as there is no such closed trail.

Our history of graph theory properly begins with its first recorded result.

Theorem 1. [Euler]. A graph is eulerian if and only if it is connected and every point has even degree.

There is another theorem due to Euler which is the basis for the statement that he is the father of topology. The formula is so classical that we do not even need to define the notation:

$$V - E + F = 2$$

for a spherical polyhedron.

2. Kirchhoff.

The next important theorem in the history of graph theory concerns the number of spanning trees of a labeled graph; this is required for calculation of currents in the branches of an electrical network. The following diagram is an example of a graph G with its spanning trees:

The graph G: has eight spanning trees, as follows:

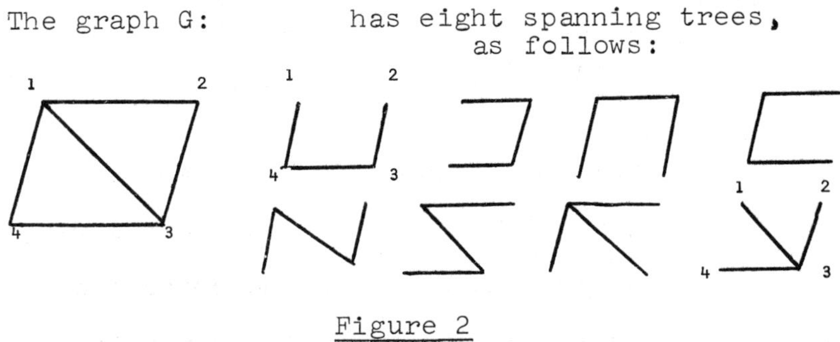

Figure 2

We illustrate Kirchhoff's derivation of the number of spanning trees for an arbitrary graph, by using the above graph G which was selected at random.

Consider the adjacency matrix A of the graph G, as labeled.

$$A = \begin{pmatrix} 0 & 1 & 1 & 1 \\ 1 & 0 & 1 & 0 \\ 1 & 1 & 0 & 1 \\ 1 & 0 & 1 & 0 \end{pmatrix}$$

Kirchhoff [11] formed from A the matrix C by taking -A and putting the degrees of the points on the diagonal.

$$C = C(G) = \begin{pmatrix} 3 & -1 & -1 & -1 \\ -1 & 2 & -1 & 0 \\ -1 & -1 & 3 & -1 \\ -1 & 0 & -1 & 2 \end{pmatrix}$$

Thus the row sums and column sums of C are all equal to zero, hence by a little-known but straightforward result from algebra, all the cofactors of this matrix are equal. Kirchhoff's result is

<u>Theorem 2</u>. [Kirchhoff]. The common value of the cofactors of C(G) is equal to the number of spanning trees of G.

Calculating the 1,1 cofactor of C, we verify that the graph G in Figure 2 has just eight spanning trees, as shown in Figure 2.

3. Cayley.

A **labeled** graph has the numbers 1,2,...,p respectively at its p points. A tree is a connected graph with no cycles. Cayley [1] was apparently the first to count labeled trees, but his result was rediscovered, generalized and independently proved many times, as pointed out most convincingly by J. W. Moon [15], who wrote an entire book on counting labeled trees.

Theorem 3. [Cayley]. The number of labeled trees with p points is p^{p-2}.

In a sense, this was already implicitly proved by Kirchhoff. For one can take the given graph as complete, namely K_p, and apply Theorem 2. Cayley's original proof was given only for p = 6, but we illustrate here using p = 5 for brevity. Consider the polynomial in five variables

$$x_1x_2x_3x_4x_5(x_1 + x_2 + x_3 + x_4 + x_5)^3.$$

It has 5^3 terms. Each of these terms corresponds to a unique spanning tree, as for example the first term

$$(x_1x_2x_3x_4x_5)x_1^3 = (x_1x_2)\ (x_1x_3)\ (x_1x_4)\ (x_1x_5).$$

The four factors on the right can be regarded as lines, and the five variables as points, giving the tree

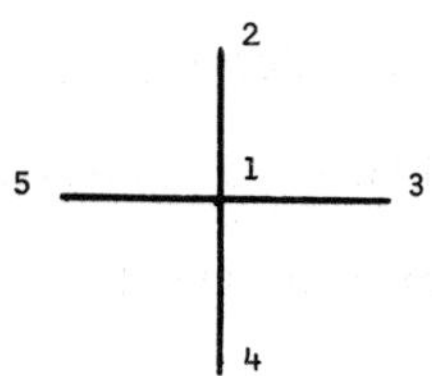

Figure 3

4. Heawood.

In addition to producing a counterexample to Kempe's "proof" of the 4CC (= Four Color Conjecture), Heawood [10] proved the 5CT (= Five Color Theorem). A *planar* graph is one that can be drawn in the plane with no pair of edges intersecting; a graph is *n-colorable* if its points can be colored using n colors in such a way that no two adjacent points have the same color.

Theorem 4. [Heawood]. Every planar graph is 5-colorable.

To this day, there have been many serious attempts to convert the 4CC into the 4CT but none has been successful as yet, and it remains a most formidable and intractible conjecture.

The next four classical theorems, in chronological order, are due to

5)	Menger	(1927)
6)	Kuratowski	(1930)
7)	Pólya	(1937)
8)	Frucht	(1938)

5. Menger.

Just as in the case of Cayley's Theorem 3, Menger's Theorem [14] has been independently rediscovered in various disguises very many times, as outlined in [7].

Theorem 5. [Menger]. In any graph G, the minimum number of points whose removal disconnects two nonadjacent points u,v is equal to the maximum number of point disjoint u-v paths.

Let $\kappa(u,v)$ denote this value. The graphs G_1 and G_2 in Figure 4 both have $\kappa(u,v) = 3$; however, in G_3, $\kappa(u,v) = 2$.

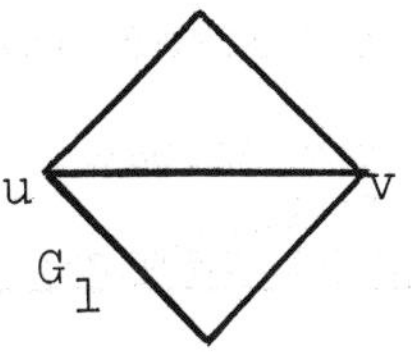

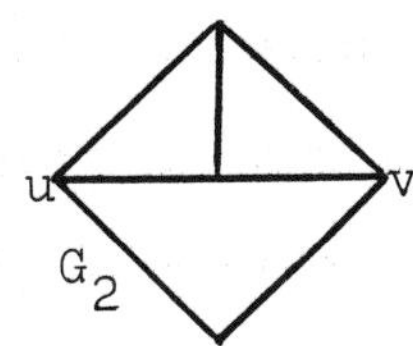

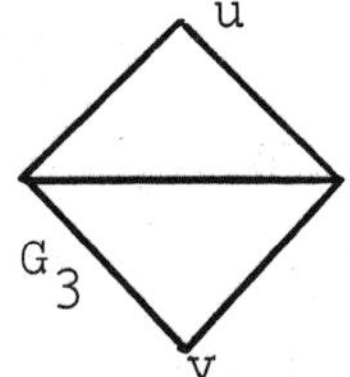

Figure 4

Dirac [2] derived from Theorem 5 the corresponding result involving line removal instead of point removal, and here u,v may be adjacent. One may prove the "max-flow, min-cut" theorem of Ford and Fulkerson [5] from Menger's theorem by forming a suitable multigraph. Then the maximum flow becomes the maximum number if line-disjoint paths, and the minimum cut becomes the minimum number of lines whose removal separates the vertices u and v.

6. Kuratowski

The determination [13] of a criterion for the planarity of a graph was a highly competitive open problem during the 1920's, as reported in the classical book, König [12]. A homeomorph of a graph is obtained by inserting additional points of degree 2 within one or more of its lines.

Theorem 6. [Kuratowski]. A graph is planar if and only if it does not contain K_5 or $K_{3,3}$ up to homeomorphism.

This was often quoted by those who looked down on our field as a theorem in graph theory whose proof is actually not obvious. The complete graph K_p has every pair of its p points adjacent; the complete bipartite graph $K_{m,n}$, also written K(m,n), has m points of one color and n points of a second color, with two points adjacent if and only if they have different colors.

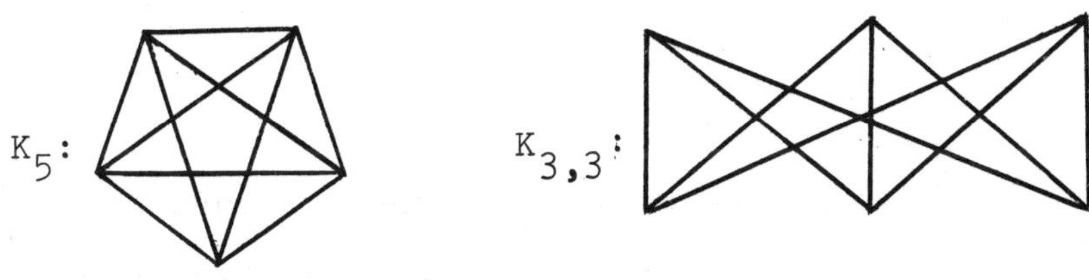

Figure 5

7. Pólya.

The most useful single result in graphical enumeration is the classical counting theorem of Pólya [17]. The principal idea is to express in a manageable form the number of orbits with specified numerical properties of an appropriate permutation group. Without explaining the notation (see [8, Chapter 15] or [9, Chapter 2]) we write c(x,y) for the "figure counting series", C(x,y) for the "configuration counting series", A for the "appropriate" permutation group, Z(A) for the "cycle index" of A and Z(A,f(x,y)) for the expression obtained from Z(A) by substituting into it the function f(x,y).

Theorem 7. [Pólya].

$$C(x,y) = Z(A, c(x,y)).$$

This formula has served to count various species of trees, graphs and many other kinds of structures.

8. Frucht.

Frucht [6] answered the question that König posed in the first book [12] ever written on graph theory. An automorphism of a graph G is an isomorphism of G with itself. Obviously the set of all automorphisms of G constitutes a group. König asked: When is a given abstract group isomorphic with the group of some graph? Frucht's reply was affirmative.

Theorem 8. [Frucht]. For every given abstract finite group F there exists a graph G such that the automorphism group of G is isomorphic to F.

To illustrate the construction used by Frucht to establish his result, we take the cyclic group of order 3. The graph for this group obtained as in Frucht's proof is

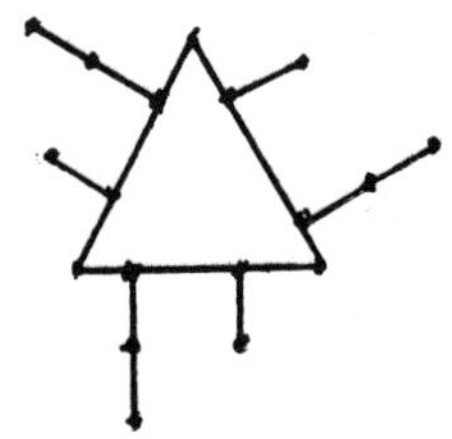

Figure 6

Clearly the group of this graph has order 3 since its only symmetries are the identity map and the rotations by $\pm \pi/3$.

The four most recent theorems in this list were established by

9)	Turán	(1941)
10)	Tutte	(1947)
11)	Nash-Williams	(1961)
12)	Ringel and Youngs	(1968)

9. Turán.

While in a German labor camp during World War II, Turán [20] discovered the pioneering theorem which initiated the study of extremal graph theory--still the most popular topic in Hungary concerning graphs. In the statement of Turán's theorem we shall use [r] to denote, as usual, the greatest integer not greater than r; similarly the symbol {r} will be used to denote the least integer not less than r.

Theorem 9. [Turán]. Among all graphs which have p points and do not contain the triangle K_3, the maximum possible number of lines is $[p^2/4]$. Further, there is a unique graph which realizes this maximum, namely the complete bipartite graph $K([p/2],\{p/2\})$.

For p = 6, the extremal graph is $K_{3,3}$; see Figure 5. More generally, Turán also derived the corresponding extremal results for graphs not containing K_n, for all $n \geq 3$, see [8, p.].

Unsolved problem. For C_4 instead of K_n exact results are not known, although upper and lower bounds have been obtained. The problem appears extremely difficult. Of course, it is also open for all larger cycles C_n.

10. Tutte.

A 1-factor, or more briefly a factor, of a graph is a spanning subgraph which is regular and of degree one. Thus it is a set of independent lines which covers all the points of the graph.

Obviously p must be even in a graph which has a factor, but this clearly is not a sufficient condition, as can be seen at once from the star $K_{1,3}$. Tutte [21] discovered the following criterion for the existence of a factor in a given graph.

Theorem 10. [Tutte]. A graph G has a factor if and only if the number of its points is even and there is no set S of points such that the number of odd components of G - S exceeds the number of points of S.

11. Nash-Williams

The *arboricity* of a graph G is the minimum number of spanning subtrees whose union is G. Alternatively, it is the smallest number of line-disjoint acyclic subgraphs (subforests) whose union is the graph. It was discovered by Nash-Williams [16]. We take Υ (capital upsilon) to denote arboricity as it is the Greek letter which most resembles a tree. As usual, let G have p points and q lines. Each tree of G has at most p - 1 lines, hence the integer

$$\{\frac{q}{p-1}\}$$

is a lower bound for the arboricity of G.

Theorem 11. [Nash-Williams]. The exact value of the arboricity of a graph G is given by

$$T(G) = \max_H \left\{\frac{q(H)}{p(H)-1}\right\} ,$$

where H is a nontrivial subgraph of G while p(H) and q(H) are the number of points and of lines of H.

It is easy to verify in two ways that

$$T(K_5) = \left\{\frac{10}{5 - 1}\right\} = 3$$

by displaying either three spanning subtrees or three line-disjoint subforests whose union is K_5.

12. Ringel and Youngs

When Heawood proved the 5CT, he also investigated the chromatic number of all other orientable surfaces S_n, the sphere with n "handles", where $n \geq 1$. This number, denoted by $\chi(S_n)$, is defined as the maximum chromatic number among all graphs which can be embedded on S_n. Heawood established the upper bound

$$\chi(S_n) \leq \left[\frac{7 + \sqrt{1+48n}}{2}\right] \qquad (n > 0),$$

and conjectured that equality always holds.

In a veritable tour de force of mathematical research, Ringel and Youngs [18,19] proved Heawood's conjecture in dual formulation by obtaining the exact value of $\gamma(K_p)$, the <u>genus</u> of K_p, that is, the minimum number n such that K_p can be embedded on the surface S_n with no pair of edges intersecting.

<u>Theorem 12</u>. [Ringel and Youngs]. The genus of the complete graph K_p is given by:

$$\gamma(K_p) = \left\{\frac{(p-3)(p-4)}{12}\right\}.$$

Heawood's upper bound for $\chi(S_n)$ becomes a lower bound for $\gamma(K_p)$, and follows readily from Euler's classical equation

$$V - E + F = 2 - 2\gamma$$

for polyhedra of genus γ. The proof that this expression is also an upper bound for $\gamma(K_p)$ could be obtained only by exhibiting an explicit embedding of K_p in the corresponding surface S_n, and was a most formidable task.

ACKNOWLEDGEMENTS

This was the first lecture at the Third Ann Arbor Conference on Graph Theory in October, 1971, and also the first of six Colloquium lectures at the July, 1972 quadrennial meeting in St. Andrews of the Edinburgh Mathematical Society. Dr. Harry Retkin of the Hatfield Polytechnic kindly assisted in the preparation of this work.

REFERENCES

[1] A. Cayley, A theorem on trees, Quart. J. Math. 23 (1889), 376-378.

[2] G. A. Dirac, Extensions of Menger's Theorem, J. London Math. Soc. 38 (1963), 148-161.

[3] L. Euler, Solutio problematis ad geometriam situs pertinentis, Comment. Academiae Sci. I. Petropolitanae 8 (1736), 128-140.

[4] L. Euler, The Königsberg bridges, Sci. Amer. 189 (1953), 66-70.

[5] L. R. Ford and D. R. Fulkerson, Maximal flow through a network, Canad. J. Math. 8 (1956), 399-404.

[6] R. Frucht, Herstellung von Graphen mit vorgegebener abstrakten Gruppe, Compositio Math. 6 (1938), 239-250.

[7] F. Harary, Variations on a theorem by Menger. Studies in Applied Mathematics 4, SIAM (1970), 112-121.

[8] F. Harary, Graph Theory, Addison-Wesley, Reading, 1969.

[9] F. Harary and E. M. Palmer, Graphical Enumeration, Academic Press, New York, 1973, to appear.

[10] P. J. Heawood, Map color theorem, Quart J. Math. 24 (1890), 332-338.

[11] G. Kirchhoff, Über die Auflösung der Gleichungen, auf welche man bei der Untersuchung der linearen Verteilung galvanischer Ströme geführt wird, Ann. Phys. Chem. 72 (1847), 497-508.

[12] D. König, Theorie der endlichen und unendlichen Graphen, Leipzig, 1936; Reprinted Chelsea, New York, 1950.

[13] K. Kuratowski, Sur le problème des courbes gauches en topologie, Fund. Math. 15 (1930), 271-283.

[14] K. Menger, Zur allgemeinen Kurventheorie, Fund. Math. 10 (1927), 96-115.

[15] J. W. Moon, Counting Labelled Trees, Canad. Math. Congress, Montreal, 1970.

[16] C. St. J.A. Nash-Williams, Edge-disjoint spanning trees of finite graphs, J. London Math. Soc. 36 (1961), 445-450.

[17] G. Pólya, Kombinatorische Anzahlbestimmungen für Gruppen, Graphen und chemische Verbindungen, Acta Math. 68 (1937), 145-254.

[18] G. Ringel and J.W.T. Youngs, Solution of the Heawood map-coloring problem, Proc. Nat. Acad. Sci. USA 60 (1968), 438-445.

[19] G. Ringel and J.W.T. Youngs, Remarks on the Heawood Conjecture, Proof Techniques in Graph Theory (F. Harary, ed.), Academic Press, New York, 1969, 133-138.

[20] P. Turán, Eine Extremalaufgabe aus der Graphentheorie, Mat. Fiz. Lapok 48 (1941), 436-452.

[21] W. T. Tutte, The factorization of linear graphs, J. London Math. Soc. 22 (1947), 107-111.

ON GRAPHS WITH A CONSTANT LINK

Morton Brown and Robert Connelly

1. Introduction.

In the Proceedings of the Symposium which took place in Smolenice, Czechoslavakia in June 1963, [5, p. 164], Zykov poses the following two problems (we shall take liberties and change his notations to ours): "Let us consider non-oriented graphs without loops and spindles; let link(x,G) be the subgraph of G generated by its vertices adjacent to x, (the link of x in G),

(a) For what graphs L is there a graph G such that all link(x,G) are isomorphic to L?

(b) For what graphs L is there only infinite graphs G with link(x,G) isomorphic to L?"

Although we do not propose to give a complete solution to either (a) or (b), we do hope to give an insight into the nature of the problems involved and provide the solutions to (a) and (b) for certain graphs L which are not too complicated. Namely, when L is the disjoint union of arcs of any length, when L is an m-ad $m \geq 3$ (a tree with only one vertex of degree greater than two), a circle of any length, and a few others. Also we shall give some general theorems relating to problem (a) on both the impossibility and possibility of building a graph G for a given L. As to problem (b), as far as we know of the history of the problem, it is not known (until now) whether there even exists one finite graph L for which there are only infinite graphs G such that link(x,G) is isomorphic to L. We propose to exhibit an infinite number of isomorphism classes of such L's. Perhaps the simplest is the graph in which $L = P_2 \cup 2P_3$ with 3 connected components, where P_n is the path with n vertices.

In fact, let us define an arc of length n as a graph with n+1 vertices $x_0, x_1, \ldots, x_n$ where there is an edge from x_{i-1} to x_i for $i = 1,2,\ldots,n$. Then we have:

<u>Theorem 1</u>. Let L be a finite disjoint union of arcs, where λ_i, $i = 1,2,\ldots,$ of the arcs have length i. Then there is a finite graph G with each link(x,G) isomorphic to L if and only if

$$\lambda_2 \leq \lambda_1 + \sum_{i=4}^{\infty} (i-3)\lambda_i = \lambda_1 + \lambda_4 + 2\lambda_5 + \ldots .$$

If we are interested in the case when G is not necessarily finite, we get a similar result except that the inequality is replaced by the conditions that when $\lambda_2 \neq 0$ either $\lambda_1 \neq 0$, $\lambda_4 \neq 0$, $\lambda_5 \neq 0$, etc. Thus we see that the L mentioned above has λ_i's in the difference between the two sets of λ_i's, and thus is an example of problem (b).

We also investigate the problem for more complicated graphs and get a similar but more complicated solution. We define an m-ad, $m \geq 3$, as a graph which is the union of m arcs identifying one endpoint of each arc to a single point. (A tree with only one vertex of degree greater than or equal to three.) Similar to Theorem 1 we have:

Theorem 2. Let L be a finite m-ad, $m \geq 3$, where λ_i, $i = 1,2,\ldots$, of the arms of L have length i. Then there is a finite graph G such that each link(x,G) is isomorphic to L if and only if

$$\lambda_2 \leq \lambda_1 + \sum_{i=4}^{\infty} (i-3)\lambda_i = \lambda_1 + \lambda_4 + 2\lambda_5 + \ldots$$

$$\lambda_1 \leq \sum_{i=2}^{\infty} \lambda_i = \lambda_2 + \lambda_3 + \ldots$$

$$2\lambda_1 \leq 2\lambda_2 + 3\lambda_3 + \sum_{i=4}^{\infty} (i-3)\lambda_i = 2\lambda_2 + 3\lambda_3 + \lambda_4 + 2\lambda_5 + \ldots .$$

Again if we are interested in the case when G is not necessarily finite then we get a similar result except the inequalities are replaced by the conditions:

$$\lambda_1 \leq \sum_{i=2}^{\infty} \lambda_i = \lambda_2 + \lambda_3 + \lambda_4 + \ldots \text{ and}$$

$$\lambda_i \neq 0 \text{ for some } i \geq 6$$

or

$$\lambda_1 \leq \lambda_2 \text{ and } \begin{cases} \lambda_2 \leq \lambda_1 + \lambda_4 \\ \lambda_5 \neq 0 \end{cases}$$

or

$$\lambda_2 \leq \lambda_1 \text{ and } \begin{cases} \lambda_1 < \lambda_2 + \lambda_3 + \lambda_4 + \lambda_5 \\ \quad \text{or} \\ \lambda_3 \neq 0 \\ \quad \text{or} \\ \lambda_4 = 0. \end{cases}$$

Thus again any set of λ_i's which is in the difference of these two sets describes an L which is an example of problem (b). An example is given by the graph L obtained by taking 3 copies of P_3 and one of P_6, and then identifying an endpoint of each of these four paths; this L at least has the virtue of being connected.

The methods and ideas employed are inspired both from combinatorial (or piece-wise linear) topology and from standard graph theory. In general the methods are quite elementary, and very little knowledge of combinatorial topology or graph theory will be either needed or assumed. However, it would be helpful if the reader were acquainted with the notions of a simplicial complex and the Euler characteristic from topology, and the König-Hall theorem, see [1], for finding the minimum flow through a transportation network. (This last prerequisite is only needed for the discussion of m-ad's however, and in many cases can be avoided even there.)

After a brief discussion of definitions, Section 3 provides the basic tool, the identification procedure, which is used throughout the rest of the paper. This is basic to almost all of the constructions to follow.

Section 4 sets up the basic building block for the construction of the more complicated graphs. Namely, it shows how to build graphs where the link of each vertex is an arc, and we are able to control the proportions of vertices with links of various lengths.

Thus in Section 4, we shall see at least how to build graphs with an arc of length not equal to 2 as constant link. In Section 4 Addendum, we apply a slightly altered building technique to build graphs, G, where each link (x,G) is isomorphic to a circle on a certain subdivision of an arbitrary graph. The completion of the proofs of Theorems 1 and 2 will appear in [12].

In most of the sections, we have tried to separate the situations for what goes into Theorem 1 and what goes into Theorem 2, even though the techniques are quite similar.

Finally, we alert the reader to the fact that in the case L = arc (path) of length ≥ 8 (path P_n with $n \geq 9$) and L = circle (cycle C_n) of length ≥ 7, our constructions run counter to non-existence theorems claimed in [11]. We believe that the error in that paper lies in the induction step of the lemma.

2. Definitions and Background.

For us, graphs will be unoriented without multiple edges or loops unless stated otherwise. As in the introduction, link (x,G) will be the subgraph determined by all the vertices adjacent to x (not including x), where x is a vertex of G. Following [7], we shall say a graph G is _Zykov regular_ (or briefly Z-regular) <u>with common link L</u> if each link(x,G) is isomorphic to L. The link(x,G), even if G is not Z-regular, will be referred to as the link of x in G, where x is a vertex of G, which as we see in a moment is motivated by the similar notion in combinatorial topology.

Basic to most of what follows is the idea of associating a simplicial complex to each graph G (as above, G is a graph as in [2] and so must have no loops or multiple edges). To each graph, G, we associate a simplicial complex, K(G), as follows: If $v_0, v_1, \ldots, v_n$ are vertices of G, and if for all i,j, $i \neq j$ an edge joining v_i and v_j is in G, then we consider a simplex, $\langle v_0, \ldots, v_n \rangle$, with those vertices as being in K(G), i.e., $\langle v_0, \ldots, v_n \rangle$ is a simplex of K(G) iff the complete graph on $v_0, \ldots, v_n$ is a subgraph of K(G). It is easy to check that this defines a simplicial complex, K(G), whose 1-skeleton is G. (See [3] or [4] for definitions and notation.)

Recall from combinatorial topology that if K is a simplicial complex, then link(x,K) = $\{\sigma \in K \mid x \cdot \sigma \in K\}$ = the set of simplices which link x. Thus, it is clear that link(x,K(G))=K(link(x,G)), and, thus, the 1-skeleton of link(x,K(G)) is link(x,g).

In what follows, we shall need certain basic graphs such as arcs, circles, and m-ads. A graph L is an <u>arc of length n</u> if L consists of n+1 vertices, $v_0,\dots,v_n$, where there is an edge between v_{i-1} and v_i for i = 1,...,n and no other edges. (Note n is the number of <u>edges</u> in L.) A graph, L, is a <u>circle of length n</u> if L consists of n vertices, $v_1,v_2,\dots,v_n$, where there is an edge joining v_i and v_{i+1} for i = 1,2,...,n-1, and an edge joining v_n and v_1, and no other edges. For an arc, the vertices v_0 and v_n are called the <u>end-points</u> of L. For $m \geq 3$, we say a graph L is an <u>m-ad of type</u> $(n_1,\dots,n_m)$, where each n_j is a positive integer, if $L = \bigcup_{j=1}^{m} L_j$, where each L_j is an arc of length n_j, and $\bigcap_j L_j$ is a single point which is an end point of each L_j. Some examples are given in Figure 1.

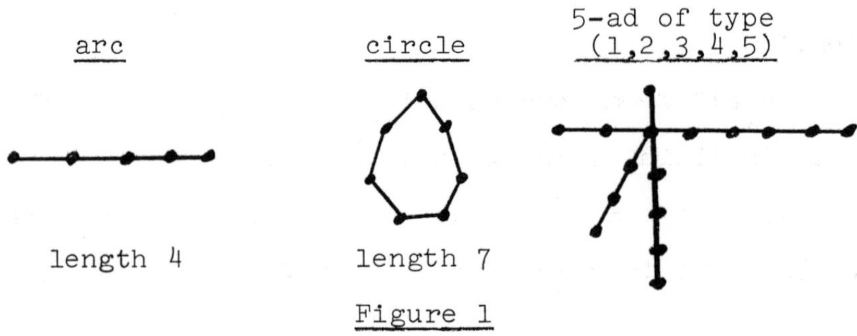

Figure 1

Note that up to isomorphisms the type of an m-ad does not depend on the ordering of integers n_j, and that a circle graph is "homogeneous" (i.e., there are graph automorphisms which take any point onto any other point) so the labeling of the vertices is not unique. Similarly, although the arc of length 1 is the only homogeneous arc, the vertices v_0 and v_n in an arc can be interchanged by a graph automorphism (which is unique), but, nevertheless, the concept of an endpoint is well defined up to isomorphism. Each of the L_j's in the definition of an m-ad above will be referred to as an <u>arm</u> of the m-ad. Note that we do not define the concept of a 1-ad or 2-ad so as not to confuse the notion of an m-ad and an arc.

3. <u>The Basic Identification Technique for Building Graphs</u>.

Most of the graphs that we build -- not just the Z-regular graphs -- can be built from simpler graphs by identifying along certain subgraphs. Thus suppose we have the following situation: Let $H_0 \subset H$ be graphs with H_0 a subgraph of H. Similarly, let $H'_0 \subset H'$ be another pair of graphs disjoint from H. Let $\phi : H_0 \to H'_0$ be a graph isomorphism, so we have the following diagram:

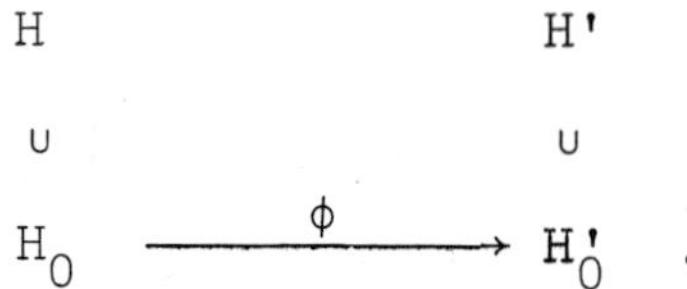

Denote by $H \underset{\phi}{\cup} H'$ the graph obtained by identifying H_0 and H_0' by ϕ. I.e., $v \sim \phi(v)$, v vertex of H_0, generates an equivalence relation on $H \cup H'$. The equivalence class of vertices are the vertices of $H \underset{\phi}{\cup} H'$. If $[v]$ and $[w]$ are two equivalence classes of vertices in $H \cup H'$, then we define an edge joining $[v]$ and $[w]$ iff there is an edge joining one of v, $\phi^{-1}(v), \phi(v)$ with one of w, $\phi^{-1}(w), \phi(w)$ (whichever is defined). This defines the graph $H \underset{\phi}{\cup} H'$ with H and H' as "subgraphs." If v or e is a vertex or edge in $H \cup H'$ let us denote by $\bar{v}$ and $\bar{e}$ the image of v and e under the identification. Similarly, if L is a subgraph of $H \cup H'$, $\bar{L}$ will denote the subgraph of $H \underset{\phi}{\cup} H'$ which is the image of the vertices and edges of L under the identification. Thus, for instance, $\bar{H} \cap \bar{H}' = \bar{H}_0 = \bar{H}_0'$.

We now relate the above process to links. In the above situation, we call an identification ϕ, <u>true</u>, iff for all vertices $v \in H$,

(i) $\overline{\mathrm{link}(v,H)} \cap \overline{\mathrm{link}(\phi(v),H')} = \overline{\mathrm{link}(v,H_0)} = \mathrm{link}(\bar{v},\bar{H}_0)$

and

(ii) $\overline{\mathrm{link}(v,H)} \cup \overline{\mathrm{link}(\phi(v),H')} = \mathrm{link}(\bar{v},H \underset{\phi}{\cup} H')$

and similarly for vertices $v \in H'$, with ϕ^{-1} replacing ϕ, and H and H' interchanged.

(If $v \notin H_0$ (or H'_0), we say $\phi(v)=\emptyset$ (or $\phi^{-1}(v) = \emptyset$)).

Remark. If ϕ is true, then it is easy to check that $\mathrm{link}(\bar{v},H \underset{\phi}{\cup} H')$ is isomorphic to $\mathrm{link}(v,H) \underset{\theta}{\cup} \mathrm{link}(\phi(v),H')$, where $\theta = \phi|\mathrm{link}(v,H_0):\mathrm{link}(v,H_0) \to \mathrm{link}(\phi(v),H'_0)$.

Thus, it is possible to compute the links of vertices in the quotient graph, $H \underset{\phi}{\cup} H'$, if we know that ϕ is true.

Also, half of (i) and (ii) automatically holds for any identification (even if ϕ is not true). Namely,

(i)' $\overline{\mathrm{link}(v,H)} \cap \overline{\mathrm{link}(\phi(v),H')} \supset \overline{\mathrm{link}(v,H_0)}$
$= \mathrm{link}(\bar{v},\bar{H}_0)$

(ii)' $\mathrm{link}(v,H) \cup \mathrm{link}(\phi(v),H') \subset \mathrm{link}(\bar{v},H \underset{\phi}{\cup} H')$

always holds for any identification. We can say,

too, that (ii) always holds for vertices. For if $w \in \mathrm{link}(\bar{v}, H \underset{\phi}{\cup} H')$ is a vertex, then there is a vertex $w' \in H$ or $w'' \in H'$ such that $\overline{\langle v, w' \rangle} = \langle \bar{v}, w \rangle$ or $\overline{\langle \phi(v), w'' \rangle} = \langle \bar{v}, w \rangle$, and in either case, $w \in \overline{\mathrm{link}(v,H)} \cup \overline{\mathrm{link}(\phi(v),H')}$.

We should also remark that it is possible for an identification not to be true even though (i) holds.

The following is an example of this phenomenon:

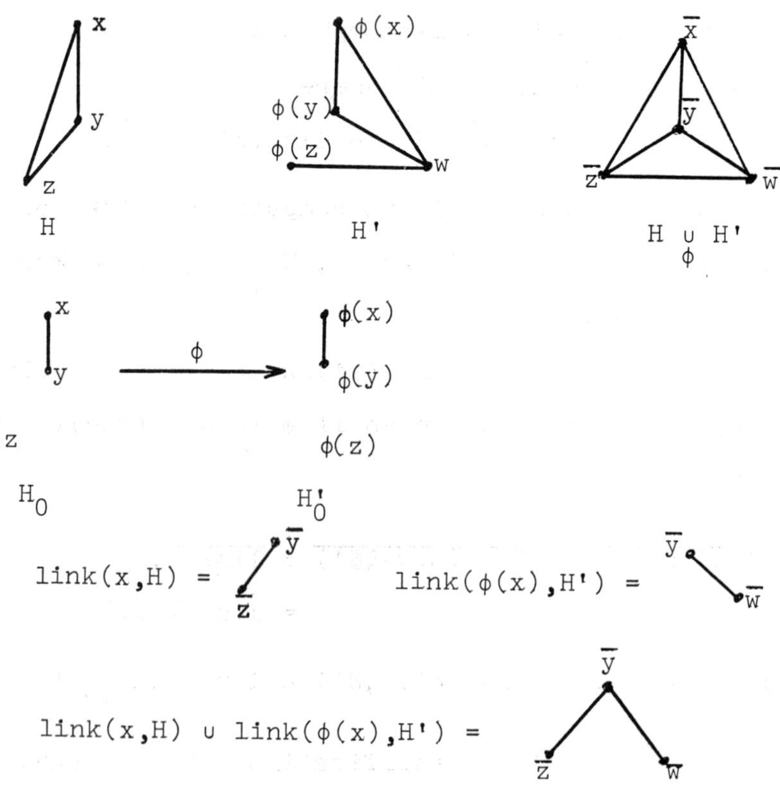

but $\text{link}(\bar{x}, H \underset{\phi}{\cup} H') =$

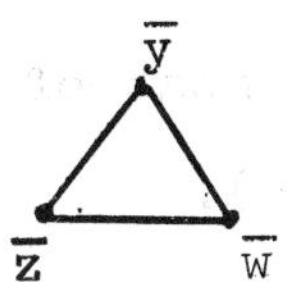

In order to build the graphs that we shall use, we need, in general, to put some conditions on how H_0 is embedded in H. We state the following definition:

If H_0 is a subgraph of a graph H, we say H_0 is <u>well-situated</u> in H if when two vertices $x, y \in H_0$ are such that

(a) x and y are in the same component of H_0

and (b) x and y are end points of an arc of length ≤ 2,

then any arc of minimal length with x and y as end points is in H_0.

For instance, if H_0 is such that any shortest path between two points in the same component of H_0 is always in H_0, then H_0 is well situated in H.

Examples of when H_0 is well situated in H abound. For instance, when H_0 consists of the union of disjoint edges (and their vertices) and vertices.

We now can state our basic lemma used for the production of most of our graphs (H, H_0, H', H_0', and ϕ are as above).

Lemma 1. Let H_0 be well situated in H. Suppose that whenever C_1 and C_2 are two different components of H_0 and are in the same component of H, then $\phi(C_1)$ and $\phi(C_2)$ are in different components of H'. Then ϕ is true.

Note that the hypothesis is not symmetric, i.e., we do not say that H_0' is well situated in H'.

Proof. By remarks above we need only to show (i)' and (ii)' for the containment reversed, and (ii) only for edges.

First we show (i) for vertices. It suffices to show (i) only for vertices $v \in H_0$, since if $v \notin H_0 \cup H_0'$, then both sides of (i) are empty. Let $w \in \overline{\text{link}(v,H)} \cap \overline{\text{link}(\phi(v),H')}$ be a vertex. Thus there are vertices $w' \in H$ and $w'' \in H'$ such that $\langle v,w'\rangle$ and $\langle \phi(v),w''\rangle$ are edges in $H \cup H'$, and $\overline{\langle v,w'\rangle} = \langle \overline{v},w\rangle = \overline{\langle v,w''\rangle}$, and since $\overline{w'} = w = \overline{w''}$, $\phi(w') = w''$. Thus if v and w' are in different components of H_0, then $\phi(v)$ and $\phi(w') = w''$ would be in different components of H'; a contradiction. Thus v and w' are in the same component of H_0, and since

H_0 is well situated in H, $\langle v,w'\rangle \in H_0$, and $w' \in \mathrm{link}(v,H_0)$, and $w \in \overline{\mathrm{link}(v,H_0)}$, and (i) holds for vertices.

Next, we show (i) for edges, and again we may assume $v \in H_0$. Let $\langle w_1,w_2\rangle \in \overline{\mathrm{link}(v,H)} \cap \overline{\mathrm{link}(\phi(v),H')}$ be an edge. By the above, there are vertices $w_1',w_2' \in H_0$ such that $\langle v,w_1'\rangle, \langle v,w_2'\rangle \in H_0$ are edges, and $\overline{w}_1' = w_1$, and $\overline{w}_2' = w_2$. Since $\langle w_1,w_2\rangle \in \overline{\mathrm{link}(v,H)}$, $\langle w_1',w_2'\rangle$ is an edge in H. Again, since H_0 is well situated in H, and w_1' and w_2' are in the same component of H_0, $\langle w_1',w_2'\rangle \in H_0$. Thus $\langle w_1,w_2\rangle \in \overline{\mathrm{link}(v,H_0)}$, and we know (ii) holds for edges.

Lastly, we show (ii) holds for edges. Let $\langle w_1,w_2\rangle \in \mathrm{link}(\overline{v}, H \underset{\phi}{\cup} H')$ be an edge. Let σ be the triangle (2-simplex) in $H \underset{\phi}{\cup} H'$ determined by the edges $\langle v,w_1\rangle, \langle v,w_2\rangle$, and $\langle w_1,w_2\rangle$. If there is a triangle in $H \cup H'$ mapped isomorphically onto σ by the identification map, then there is a $\langle w_1',w_2'\rangle \in H \cup H'$, an edge, such that $\overline{\langle w_1',w_2'\rangle} = \langle w_1,w_2\rangle$, and $\langle w_1',w_2'\rangle \in \mathrm{link}(v,H) \cup \mathrm{link}(\phi(v),H')$, and (ii) holds. Otherwise, two of the edges of σ have a preimage in one of H or H' and the third edge a preimage in the other. Let **e** be the third edge in σ where $e = \langle x_1,x_2\rangle$, and $e' = \langle x_1',x_2'\rangle$, a preimage in

$H \cup H'$. If $e' \varepsilon H_0 \cup H_0'$, then $\phi(e')$ or $\phi^{-1}(e')$, whichever is defined, completes the third side of the triangle in the preimage of the other two sides of σ, and we are in the previous case. If $e' \not\in H_0 \cup H_0'$, and no triangle is in the preimage of σ, then both $x_1', x_2' \varepsilon H_0 \cup H_0'$. Suppose $e' \varepsilon H$. Since H_0 is well situated in H, x_1' and x_2' must be in different components of H_0, but being connected by e', by the hypothesis on ϕ, $\phi(x_1')$ and $\phi(x_2')$ must be in different components of H'. But $\phi(x_1')$ and $\phi(x_2')$ are connected in H' by the preimage of the other two sides of σ. Thus we have a contradiction. Suppose $e' \varepsilon H'$. Clearly $<\phi^{-1}(x_1'), \phi^{-1}(x_2')>$ is not an edge in H, otherwise a triangle in H maps onto σ. Let e_1, e_2 denote the preimage of the other two sides of σ in H. If $\phi^{-1}(x_1')$ and $\phi^{-1}(x_2')$ are in the same component of H_0, then since H_0 is well situated in H, $e_1, e_2 \varepsilon H_0$, and there is a triangle in H', $(\phi(e_1), \phi(e_2), e')$, mapped isomorphically onto σ, a contradiction. If $\phi^{-1}(x_1')$ and $\phi^{-1}(x_2')$ are in different components of H_0, then $\phi\phi^{-1}(x_1') = x_1'$ and $\phi\phi^{-1}(x_2') = x_2'$ must be in different components of H'. But they are connected by e', a contradiction. Thus in all cases (ii) holds.

Corollary 1. Let H_0 be the disjoint union of edges (and their vertices of course) and single vertices. Suppose that whenever C_1 and C_2 are two different components of H_0 and are in the same component of H, then $\phi(C_1)$ and $\phi(C_2)$ are in different components of H'. Then ϕ is true.

Let A_n be the following graph, known as annulus of length n.

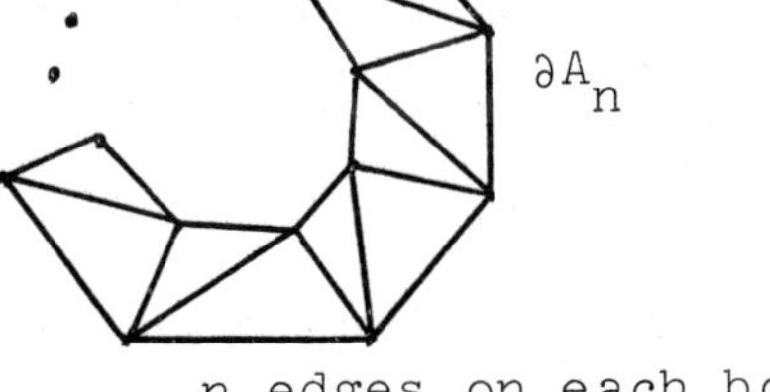

Figure 2

Let ∂A_n denote the subgraph consisting of the 2n edges and vertices on the boundary as indicated.

Corollary 2. Let H be the disjoint union of annuli as above (not necessarily of the same length) with H_0 the union of the boundaries of each component of H. Let ϕ be as before with components. Then ϕ is true.

Proof. H_0 is well situated in H.

4. Building Graphs with Arcs as Links.

In order to build more complicated graphs, we must first understand how to build certain graphs where the link of each vertex is an arc (of possibly varying length). Also, in passing, we shall show how to build finite Z-regular graphs, where the common link is an arc of length not equal to 2.

Throughout this section, we shall only consider those graphs H which are finite and each link(v,H) is an arc of non-zero length. If H is such a graph, we define ∂H to be the subgraph of H consisting of the vertices of H together with those edges $\langle v,w \rangle \in H$ such that w is an endpoint in the link(v,H) (and thus v is an endpoint in link(w,H), also). Note that $K(\partial H)$ is the ordinary manifold boundary of the 2-manifold K(H) in combinatorial topology. In any case, it is easy to see that ∂H consists of the disjoint union of circles. We shall say a graph H has <u>type</u> $(n_1, n_2 \ldots n_k)$, where each n_i is a positive integer, if we can write the vertices, $v_1, v_2 \ldots v_n$, of each component of ∂H such that

(a) $\langle v_1, v_n \rangle, \langle v_i, v_{i+1} \rangle$ $i = 1, \ldots n-1$ are edges in ∂H and

(b) $n = k\ell$, ℓ an integer, and $\text{link}(v_{jk+i}, H)$ has length n_i, where $j = 0,1,\ldots \ell-1$, $i = 1,2 \ldots k$.

In other words the vertices of each ∂H have links of length n_1 then n_2, etc., cyclicly repeated as one proceeds around ∂H.

Note that not all graphs have a type, and that a type is not necessarily unique -- even up to cyclic permutation or reversal. We are more interested in given a type, a sequence of integers, whether some graph has that type. In fact, if a graph has a type $(n,n, \ldots,n)$, then it is a Z-regular graph whose common link has length n. From here on, it is very handy to briefly say $(n_1,n_2 \ldots n_k)$ exists to mean that there is some graph H such that H has a type $(n_1,n_2 \ldots n_k)$. Also, we use the notation $(n_1,n_2,\ldots,n_k,n_1,n_2 \ldots n_k, n_1 \ldots n_k,\ldots) = (n_1,\ldots,n_k)^j$, where $n_1,\ldots,n_k$ is repeated j times. We can now state some observations where all but (2) are immediate: For $k \geq 2$,

(0) $\exists (n_1,n_2,n_3 \ldots n_k) \Longleftrightarrow \exists (n_2,n_3,\ldots,n_k,n_1) \Longleftrightarrow \exists (n_k,n_{k-1},\ldots n_1)$

(1) $\exists(n_1,n_2,\ldots,n_k)^j \Longrightarrow \exists(n_1,\ldots,n_k)$

(2) $\exists(m_1,m_2 \ldots m_\ell)$ and $\exists(n_1,n_2,\ldots,n_k)$

$\Longrightarrow \exists(m_1+n_1,m_2,m_3,\ldots,m_{\ell-1},m_\ell+n_2,n_3,\ldots,n_k)$

$(\ell,k \geq 2)$

(3) $\exists(1,1,1)$.

Proof of (0),(1),(3). (0) and (1) follow directly from the definition of a type. (3) follows from the existence of the triangle graph, K_3.

Proof of (2). The technique of proof here will be quite typical of the identification techniques to follow. Let H_1 be a graph with a type $(m_1,\ldots,m_\ell)$ and H_1' a graph with a type $(n_1,\ldots,n_k)$. Take a fixed labeling of each vertex in H_1 and H_1', where the vertices are labeled in order as we travel around each component of ∂H_1(or $\partial H_1'$), and, if a vertex, v, is labeled i, then link(v,H_1) (or link(v,H_1')) is an arc of length m_i (or n_i). Consider the edges in ∂H_1 that join a vertex labeled 1 with a vertex labeled ℓ. Suppose there are p of them, $e_1,e_2 \ldots e_p$. Similarly, consider the edges in $\partial H'$ that join a vertex labeled 1 with a vertex labeled 2. Suppose there are q of them,

$e'_1,\ldots,e'_q$. Let H be q disjoint copies of H_1 and H' p disjoint copies of H'_1. Let $e_{i,j}$, $i = 1,2 \ldots p$, $j = 1,2 \ldots q$, denote the edge e_i in the jth copy of H_1 in H. Similarly, let $e'_{i,j}$, $i = 1,\ldots q$, $j = 1,\ldots,p$, denote the edge e'_i in the jth copy of H'_1 in H'. Let H_0 be the graph determined $\underset{i,j}{\cup} e_{i,j}$ and H'_0 the graph determined by $\underset{i,j}{\cup} e'_{i,j}$. Let $\phi : H_0 \to H'_0$ be a graph isomorphism such that

(a) ϕ(any vertex labeled 1) = vertex labeled 1

(b) If $e_{i_1,j_1} \neq e_{i_2,j_2}$ but are in the same component of H, then $\phi(e_{i_1,j_1})$ and $\phi(e_{i_2,j_2})$ are in different components of H'.

Such a ϕ can be easily defined by saying $\phi(e_{i,j}) = e'_{j,i}$, and then ϕ is completely defined by condition (a).

We now claim the graph $H \underset{\phi}{\cup} H'$ has a type $(m_1+n_1,m_2 \ldots m_\ell + n_2,n_3 \ldots n_k)$. First, by conditions (b) and Corollary 2 of Section 3, ϕ is true. Thus the link of any vertex not in H_0 or H'_0 has not changed, and by the remark after the definition of true, if v is say one of the vertices labeled 1 in

H (or H'), then link($\overline{v}$, $H \underset{\phi}{\cup} H'$) is an arc of length $m_1 + n_1$. Similarly, for a vertex w, labeled ℓ in H (as 2 in H'), the link ($\overline{w}$, $H \underset{\phi}{\cup} H'$) in an arc of length $m_\ell + n_2$. Lastly, it is easy to check that $\partial(H \underset{\phi}{\cup} H')$ consists of arcs alternating in $\overline{H}$ and $\overline{H}'$, where we first have sequence of vertices labeled $1,2,\ldots,\ell$ in $\overline{H}$ then $2,3,\ldots,k,1$ in $\overline{H}'$, where the ℓ vertex in $\overline{H}$ is the same as the 2 vertex in $\overline{H}'$, and the 1 vertex of $\overline{H}$ is the same as the 1 vertex of $\overline{H}'$. Thus $H \underset{\phi}{\cup} H$ has the appropriate type. The following diagram makes this clear.

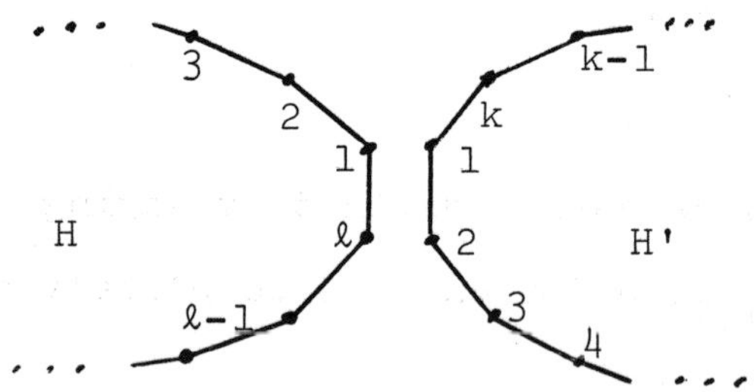

Figure 3

We now proceed to make some statements that follow easily from the statements (0),(1),(2), and (3). In fact, if we wish, we could now treat (0),...,(3) as axioms and regard the forthcoming statements as formal manipulations of (0),...,(3). However, we shall interpret the statements geometrically as well. It is interesting, however, that if we drop statement (1), then all the graphs involved (interpreting geometrically) are planar and in fact, represent triangulations of a 2 simplex. These graphs are also called maximal outerplanar graphs (see [2]) and are completely determined by their type of maximal length, which is of course unique.

(4) $\exists(n_1, n_2, \ldots, n_k)^j,\ kj > 1,$

$\Longrightarrow \exists(n_1+2, 1, n_2+2, 1, \ldots, n_k+2, 1)^j.$

<u>Proof of (4)</u>. Repeated applications of (2), kj times, where $(m_1, m_2, m_3) = (1,1,1)$ and (0) is used to shift to the appropriate spot. Geometrically, this is the process of "attaching triangles" to the boundary.

(5) $\exists(n_1, \ldots, n_k), k > 1 \Longrightarrow \exists(n_1+n_2, n_3, \ldots, n_k)^2.$

<u>Proof of (5)</u>. By (2) and (0), $\exists(n_2, n_3 \ldots n_k, n_1)$ and $\exists(n_1, n_2, \ldots, n_k) \Longrightarrow$

$\exists(n_2+n_1, n_3, \ldots, n_k, n_1+n_2, n_3, \ldots, n_k)$

$\Longleftrightarrow \exists(n_1+n_2, n_3 \ldots n_k)^2.$

(6) $\exists (n_1, n_2, \ldots, n_k)^j, kj > 1, \Longrightarrow (n_1+3, n_2+3, \ldots, n_k+3)^j.$

<u>Proof of (6)</u>. Apply (4) and (5) and (1) jk times (shifting to the appropriate spot by (0) of course).

(7) $\exists (1,2)^2$ and $\exists (1,3)^3$.

<u>Proof of (7)</u>. Apply (5) to (3) to show $\exists (2,1)^2 \Longleftrightarrow \exists (1,2)^2$ by (0).

Apply (4) to (3) to show $\exists (3,1)^3 \Longleftrightarrow \exists (1,3)^3$ by (0).

The graphs for these two types, which are unique, are:

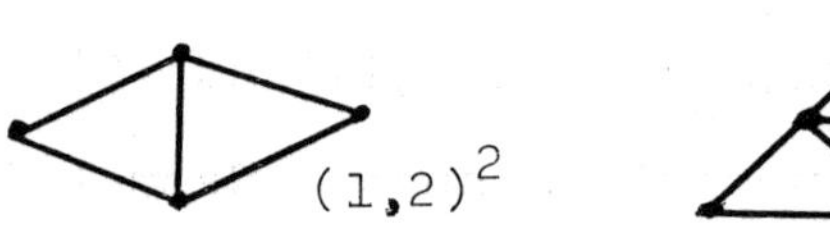

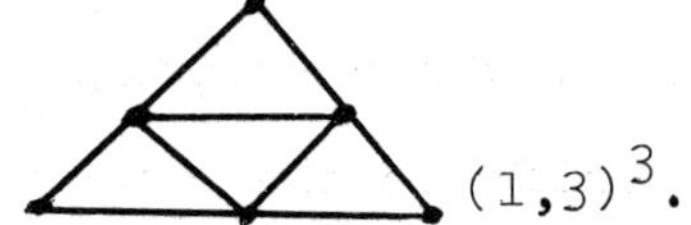

<u>Figure 4</u>

(8) $\exists (n)^2$, $n \geq 3$.

<u>Proof of (8)</u>. By (1) applied to (7) (then (0)) $\Longrightarrow \exists (1,2)$ and $\exists (2,1) \Longrightarrow \exists (3,3)$ by (2).

Similarly, $\exists (1,3)$ and $\exists (3,1) \Longrightarrow \exists (4,4)$. $\exists (1,3)$ and $\exists (2,1) \Longrightarrow \exists (3,4)$. $\exists (3,4)$ and $\exists (2,1) \Longrightarrow \exists (5,5)$.

This shows (8) for n = 3,4,5. We apply (6) (or $\exists$(3,3)) the appropriate number of times to get the general case.

Note we now have constructed a Z-regular graph with an arc of length n for every n $\geq$ 3, and further each boundary component has even length.

Note also that the graph of type (3,3) is our old friend the annulus, if one chases through the identifications.

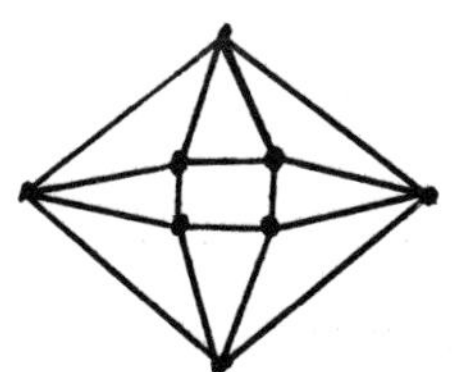

Figure 5

We remark, too, that we did not show $\exists$(2,2) (or even $\exists$(2)) because it is false, as is easily shown directly, but later we shall show $\nexists$(2) as a special case of other results.

Later, in order to build more complicated Z-regular graphs, we shall need the following:

(9) $\exists$(1,n), n $\neq$ 1,4.

Proof of (9). By (7), we assume n $\geq$ 5. Then apply (4) to (8).

(10) $\exists(\underbrace{2,2,\ldots,2}_{(n-3)\ 2\text{'s}},n)^2$, $n \geq 3$.

Proof of (10). First, we show $\exists(n,1,2,2,2 \ldots 2)$, $n \geq 2$ with (n-2) 2's by induction, $\exists(2,1)$ by (7). Then by (3), $\exists(1,1,1)$ and $\exists(n,1,\underbrace{2,2 \ldots 2}_{(n-2)2\text{'s}}) \Longrightarrow$

$\exists(n+1,1,\underbrace{2,\ldots,2}_{(n-1)\ 2\text{'s}})$ by (2). Lastly, we apply (5) and (0)).

(11) $\exists(\underbrace{1,4,1,4,\ldots,1,4}_{n-5\text{ times}},1,n)^2$ $n \geq 5$

Proof of (11). Apply (4) to (10).

(12) $\exists(1,4,1,3)$.

Proof of (12). Apply (4) to (7).

(13) $\exists(1,4,4)^2$.

Proof of (13). By (7) and (1) applied to (7) $\exists(1,2)^2$, and $\exists(2,1) \Longrightarrow \exists(3,3)^2$ by applying (2) twice. Similarly, $\exists(3,3)^2$, and $\exists(1,1,1)$ $\Longrightarrow \exists(4,1,4)^2$ by applying (2) twice (as well as

(0) of course). An example of a graph with this type is the following:

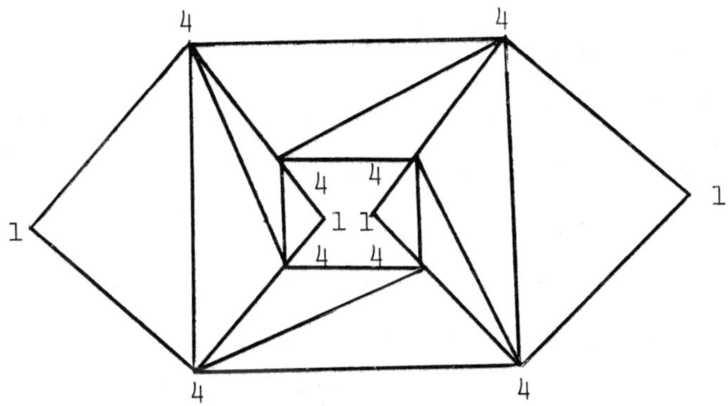

Figure 6

4, Addendum. Building Z-regular Graphs with Certain Other Graphs as Common Link.

Apart from the more complicated techniques of the preceding sections, we can describe some simple techniques which allow us to build various Z-regular graphs. To demonstrate the technique, we shall apply it first to "construct" circles.

Theorem 3. Let L be a circle with $n \geq 3$ edges. Then there is a finite Z-regular graph G with L as its common link.

Proof. If $n = 3,4$, or 5 then we can let G be a

tetrahedron (3 simplex), octahedron, or an icosahedron respectively.

If $n \geq 6$, we proceed as follows: By (8) of Section 4, there is a Z-regular graph H_1 with an arc of length n-3 as common link. Let H' denote 2 disjoint copies of H_1. Let H denote the disjoint union of annuli, one for each component of ∂H_1, and of length equal to the length of that component. Recall an annulus of length n is the graph

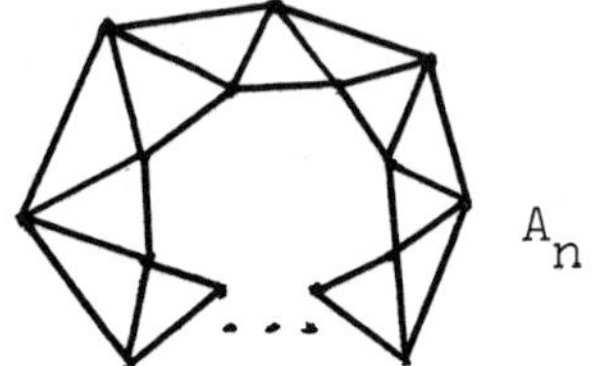

Figure 7

where there are two boundary components to A_n each of length n. (The link of every point in A_n is an arc of length 3.) Let $H_0 \subset H$ be the boundary of H, the annuli, and $H_0' \subset H'$ the boundary of H'. Let $\phi : H_0 \to H_0'$ be the isomorphism which takes each circle of H_0 to the corresponding boundary in one copy of H_1 and the other boundary circle on the annulus in H_0 to the other copy of the same boundary component in $H_0' \subset H'$. By Corollary 2, ϕ is true. Thus, $H \underset{\phi}{\cup} H'$ is a Z-regular graph with a circle of length n as common link.

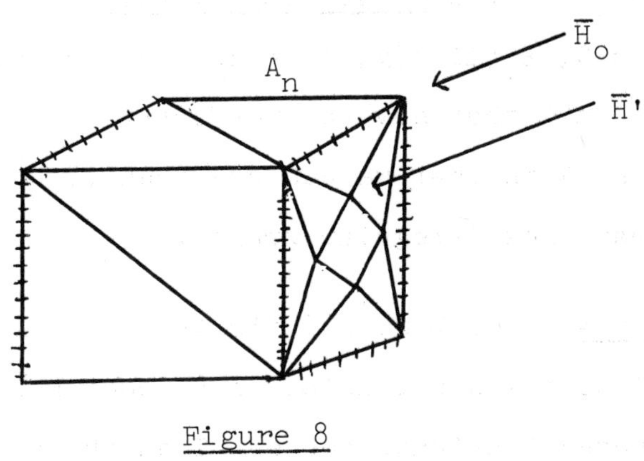

Figure 8

We shall show how to extend the above idea to more general graphs. In particular, suppose L is any graph (finite). Let us denote by L' the graph obtained by replacing each edge <v,w> by the arc of length 3 with v and w as endvertices, (L' is a subdivision of L). We shall show that for any finite graph, L,L' is the common link of a finite Z-regular graph.

Before we proceed to the proof of this, however, we need one more definition. As in Section 4, we shall regard a Z-regular graph, G, with common link, L, as the graph, G, together with a preferred collection of isomorphisms θ_x : link(x,G) → L. We must fix our attention on some one θ_x that provides the isomorphism. With this in mind, we say a one-factor $F \subset G$ <u>points to a vertex</u> $x_F \in L$, if for all $y \in G$, $\theta_y(\text{link}(y,F)) = x_F$. As we shall see, points in L that have a one-factor point at them are useful in that we can use them to build onto L to create new Z-regular graphs.

<u>Theorem 4</u>. Let L be any finite graph without loops. Let L' be the subdivision of L (adding 2 vertices to each edge) mentioned above. Then there is a finite Z-regular graph G with L' as its common link.

<u>Proof</u>. We shall prove a slightly stronger statement: Namely, there is a finite Z-regular graph, G, with L' as its common link and for each vertex, $v \in L$, regarding v as a vertex of L', there is a one-factor, $F_v \subset G$, such that F_v points to v.

We shall prove the above statement by induction on the number of edges of L. First, if L has no edges and is the discrete union of vertices $v_1, v_2, \ldots, v_n$, we may let G be the 1-skeleton of an n-cube. Alternatively, we think of G itself being build inductively (on the number of vertices) as follows: If n = 0, let G be a single vertex ($L=\emptyset$). If n = 1, let G_1 be a single edge, and $F_{v_1} = G_1$. Thus, suppose a G_{n-1} is constructed for $n-1 \geq 0$.

Let G_n be two copies of G_{n-1} with an arc joining corresponding vertices. The union of the new arcs will be F_{v_n} for G_n and the two copies of F_{v_i}, $i < n$, in G_n will be the new F_{v_i}. Thus, the statement is seen to be true when L has no edges.

Suppose $e \in L$ is an edge, $e = \langle v,w \rangle$. By the induction hypothesis, there is a finite Z-regular graph, H_1, with (L-e)' as its common link and two one-factors F_v, $F_w \subset H_1$ pointing to v and w respectively. Let $C = F_v \cup F_w$. Note C is the disjoint union of circles of even length. Let H' be two disjoint copies of H_1. Let H be the disjoint union of annuli, where there is one annulus for each component of C and of the same length as that component. Let $H_0 = \partial H$ and H_0' be the two copies of C in H'. Let $\phi : H_0 \to H_0'$ be the isomorphism which takes each circle of the boundary of each annulus into the corresponding circle in C, one component in one copy of C and the other in the other copy of C. Let $G = H \underset{\phi}{\cup} H'$. Since H_0 is well situated in H, ϕ is true; thus, the link of each vertex in G is an arc of length 3 attached to v and w in (L-e)', which is simply L'. In particular, if a vertex $x \in H$, then since ϕ is true, link$(\overline{x}$,G) is isomorphic to

$\text{link}(x,H) \underset{\phi|\text{link}(x,H_0)}{\cup} \text{link}(\phi(x),H')$ by the remark after the definition of true. And, in fact, it is easy to check that the isomorphism is such that if we call $\theta_{\bar{x}} : \text{link}(\bar{x},G) \to L \cong \text{link}(x,H) \underset{\phi|\text{link}(x,H_0)}{\cup} \text{link}(\phi(x),H')$, the isomorphism $\theta_{\bar{x}} \mid \bar{H}' = \theta_{\phi(x)} q_2^{-1}$ where $q_2 : H' \to \bar{H}'$ is the quotient. Thus, if we let F_z denote the one-factor in H' (from the one in H_1) pointing to $z \in (L-e)'$ (z a vertex in L), then $\bar{F}_z$ is a one-factor in G still pointing to z. Thus, the inductive step is proved.

Note that by the above process, if L is a graph with p vertices and q edges, then the number of vertices in G is 2^{p+q}. (Note G has L' and not L as the common link.)

Note also, that by the above process it is in general not possible to find a one-factor pointing to the other vertices of L'.

REFERENCES

[1] C. Berge, Theory of Graphs and Its Applications, Wiley, New York, 1962.

[2] F. Harary, Graph Theory, Addison-Wesley, Reading, 1969.

[3] J.F.P. Hudson, Piecewise Linear Topology, Benjamin, New York, 1969.

[4] S. Lefschetz, Introduction to Topology, Princeton, 1969.

[5] H. Seifert and W. Threlfall, Lehrbuch der Topologie, Teubner, Leipzig, 1934. reprinted, Chelsea, New York, 1947.

[6] A.A. Zykov, Problem 30, Theory of Graphs and Its Applications, Proceedings of the Symposium held in Smolenice in June, 1963, Academic Press, 1964, 164-165.

[7] H. Izbicki, On the electronic generation of regular graphs, Theory of Graphs, International Symposium, Gordon and Breach, New York, 1967, 177-182.

[8] J. Seidel, Strongly regular graphs, Indag. Math. 29 (1967), 188.

[9] H.H. Teh and P. Jha, Various constructions of regular graphs, Nanta Math. (1965), 31-44.

[10] H. Izbicki, Regular graphs of arbitrary degree having prescribed properties, (German) Monatsh Math. 64 (1960), 15-21.

[11] S. Ya. Akishieva, On graphs with given circumferences, Mat. Zament. 3 (1968), 211-216.

SOME EXTREMAL PROBLEMS ON r-GRAPHS

W. G. Brown, P. Erdös, and V. T. Sós

1. Introduction.

By an _r-graph_ we mean a fixed set of _vertices_ together with a class of unordered subsets of this fixed set, each subset containing exactly r elements and called an _r-tuple_. In the language of Berge [2] this is a _simple uniform hypergraph of rank_ r. The concept becomes interesting only for $r > 1$: for $r = 2$ we obtain ordinary (i.e., Michigan) graphs. We shall represent an r-graph by a capital Latin letter followed by a superscripted (r), as $H^{(r)}$. If in some context this symbol is followed by (n) or (n,m), as $H^{(r)}(n)$ or $H^{(r)}(n,m)$, this will mean that $H^{(r)}$ has exactly n vertices and, in the second case, _at least_ m r-tuples; thus n will always be an integer, but m need not be. Any of the foregoing notations, when applied to the symbol for a _family_ of r-graphs, will be intended to apply to all members of the family: for example, if $\mathcal{H}^{(r)}$ is a family of r-graphs, and we write $\mathcal{H}^{(r)}(n)$, we shall be saying that every member of $\mathcal{H}^{(r)}$ has exactly n vertices. The letter G will be reserved for a general r-graph: in the sense that we may write "any $G^{(r)}(n;m)$" when we mean "any r-graph having n vertices and at least m r-tuples"; it will not be used as the name of a specific r-graph. As another example, if we use the symbol $\mathcal{H}^{(r)}(n)$ defined above, we are saying that every member of the family $\mathcal{H}^{(r)}$ is a $G^{(r)}(n)$. The superscripted (r) will sometimes be omitted from the symbol for a family of r-graphs, but will normally be included in the symbol for a specific r-graph -- except possibly when r = 2.

For any fixed family $\mathcal{H}$ of r-graphs and any positive integer n, the extremal number ex(n; $\mathcal{H}$) is the largest integer t for which there exists a $G^{(r)}(n,t)$ containing no member of $\mathcal{H}$ as a sub-r-graph. More precisely, $\mathcal{H}$ is a family of isomorphism classes of r-graphs, none of which contains an r-graph, which may be extended, possibly by adjoining new vertices and/or new r-tuples, to yield this $G^{(r)}(n,t)$. In the case of graphs, r = 2, much work has been done on extremal numbers . As usual, K_t denotes the complete graph with t vertices. Turán [11] generalized a result of Mantel-Wythoff [9] by evaluating $ex(n; \{K_t^{(2)}(t,\binom{t}{2})\})$ for every t. Numerous other results, both exact and asymptotic, have since been discovered for graphs: indeed, for every family $\mathcal{H}$ of graphs, $n^{-2}ex(n; \mathcal{H})$ approaches a known limit [8] as n approaches infinity, the value of the limit depending only on the minimum of the chromatic numbers of members of $\mathcal{H}$; descriptions of general extremal results for graphs may be found in [5] etc. In [4] we investigated several problems for $r = 3$. For example, we studied the asymptotic behavior of $ex(n; \mathcal{T}^{(3)})$ where $\mathcal{T}^{(3)}$ is the class of all triangulations of the sphere -- thereby generalizing the trivial statement that any graph $G^{(2)}(n;n)$ contains a polygon.

Let $\mathcal{G}^{(r)}(n,m)$ denote the class of <u>all</u> $G^{(r)}(n,m)$. In this paper we shall denote $ex(n;\mathcal{G}^{(r)}(k,h))$ by $f^{(r)}(n;k,s) - 1$. Thus $f^{(r)}(n;k,s)$ denotes the smallest t for which every $G^{(r)}(n,t)$ contains at least one $G^{(r)}(k,s)$. This problem was studied for graphs in [5] and for $r = 3$ in [4]. In the former case, many exact values are known, as well as asymptotic information; in the latter case, there are many gaps -- even for small values of k. We shall give examples of some known results for $r = 2$ and

$r = 3$. The main result of this paper consists of the determination of a lower bound for $f^{(r)}(n;k,s)$. The method of proof is that called by one of us [6] "probabilistic" -- we employ a counting argument to prove the existence of r-graphs containing no $G^{(r)}(k,s)$ and having the desired number of r-tuples, but we make no attempt to exhibit the r-graphs explicitly. The bound we obtain is not always best possible, but does in some cases improve on our earlier results for $r = 3$.

The letter c, possibly subscripted, will be reserved for positive constants which appear in inequalities for extremal numbers. We shall not be concerned with best possible values for such constants.

2. Some known values of $f^{(2)}(n;k,s)$

We shall not attempt an exhaustive discussion here, but refer the reader to [5] remarking, however, that some of the results there stated have been improved upon by various authors. We discuss below the behavior when $s \leq k$.

First, for the range $s < k$,

$$f^{(2)}(n;k,s) = \begin{cases} s & s \leq k/2 \\ 1 + [n(2s-k)/2s-k+1] & k/2 < s < k. \end{cases}$$

When $s = k$ an exact result [9,11] is available only for $k = 3$, where

$$f^{(2)}(n;3,3) = [n^2/4] + 1.$$

Some information is available on the asymptotic behavior of $f^{(2)}(n;k,k)$. When $k = 4$, it follows from a previous result [3,7] that

$$\lim_{n\to\infty} n^{-3/2} f^{(2)}(n;4,4) = 1/2.$$

Erdös has proved the existence of positive constants ε_k, a_k, b_k such that the inequality

$$a_k n^{1+\varepsilon_k} < f^{(2)}(n;k,k) < b_k n^{1+1/[k/2]}$$

holds for all k: indeed, with $\varepsilon_k = 1/[k/2]$ for $k \leq 5$ at least. This stronger lower bound is easily seen to be valid for $k = 6,7$ and for $k = 10,11$ using graphs derived from families constructed by Benson [1] and Singleton [10].

3. Some known values of $f^{(3)}(n;k,s)$.

Much of [4] was devoted to a discussion of other structural problems. But we did determine asymptotic bounds for $f^{(3)}(n;k,s)$ for $k \leq 6$. We reproduce here the list of inequalities proved in Theorem 4 of that paper:

$$\lim_{n\to\infty} n^{-2} f^{(3)}(n;4,2) = 1/6$$

$$c_3 n^3 < f^{(3)}(n;4,3) < f^{(3)}(n;4,4)$$

$$f^{(3)}(n;5,2) = [n/3] + 1$$

$$c_4 n^2 < f^{(3)}(n;5,3) < c_5 n^2$$

$$c_6 n^{5/2} < f^{(3)}(n;5,4) < c_7 n^{5/2}$$

$$c_8 n^3 < f^{(3)}(n;5,5) < \ldots < f^{(3)}(n;5,10)$$

$$f^{(3)}(n;6,2) = 2$$

$$c_9 n^{3/2} < f^{(3)}(n;6,3)$$

$$c_{10} n^2 < f^{(3)}(n;6,4) < n^2/4$$

$$f^{(3)}(n;6,6) < c_{11} n^{5/2}$$

$$f^{(3)}(n;6,8) < c_{12} n^{11/4}$$

$$c_{13} n^3 < f^{(3)}(n;6,9) < \ldots < f^{(3)}(n;6,20)$$

Perhaps the most interesting question we were unable to answer is whether $f^{(3)}(n;6,3) = o(n^2)$. Our main result will provide improved lower bounds for $f^{(3)}(n;6,5)$ and $f^{(3)}(n;6,6)$; also we shall be able to generalize the pair of inequalities for $f^{(3)}(n;6,4)$ to $f^{(3)}(n;k,k-2)$.

4. A lower bound for $f^{(r)}(n;k,s)$.

Our main result is now stated.

Theorem. For integers $k > r$ and $s > 1$ there exists a positive constant $c_{k,s}$ such that

$$f^{(r)}(n;k,s) > c_{k,s} n^{(rs-k)/(s-1)}$$

Before proceeding with the proof, which uses the so-called "probabilistic" methods of [6], we remark that the exponent of n in the above inequality is not always best possible. It can, however, be shown to be best possible when $s - 1$ divides $rs - k$. For example, when $k = 5$ and $s = 4$ we know that

$f^{(3)}(n;5,4) = O(n^{5/2})$, but here we obtain only $f^{(3)}(n;5,4) > cn^{7/3}$.

Proof of the theorem. Let r,k,s be fixed integers ($k > r$, $s > 1$). Let n be any integer "sufficiently large", and m an integer to be further specified in inequality (1) below. Let V be a fixed set of cardinality n. The set of r-graphs having vertex set V and exactly m r-tuples will be denoted by M; it has exactly $\binom{\binom{n}{r}}{m}$ members. For any r-graph $H^{(r)}$ in

M, a subset K of V of cardinality k is called "$H^{(r)}$-bad" if at least s r-tuples of $H^{(r)}$ are contained in K. Denoting by $b(H^{(r)})$ the number of $H^{(r)}$-bad subsets of V, we shall choose m so that the following inequality will hold:

$$(1) \qquad \Sigma \; b(H^{(r)}) \div \binom{\binom{n}{r}}{m} \leq m \div 2 \binom{k}{r},$$

where the sum is taken over all $H^{(r)} \varepsilon$ M, both in (1) and below. Since the left member of this inequality is just the average number of $H^{(r)}$-bad subsets in graphs in family M, (1) ensures that there exists an r-graph $H_o^{(r)}$ in M such that $b(H_o^{(r)}) < m/\binom{k}{r}$. If we omit from $H_o^{(r)}$ every r-tuple which occurs in an $H_o^{(r)}$-bad k-tuple we may construct a $G^{(r)}(n,m)$ containing no $H_o^{(r)}$-bad k-tuple: hence it will follow that

$$f^{(r)}(n;k,s) \geq m.$$

It remains to determine for given n the largest m for which (1) holds. The total number of $H^{(r)}$-bad k-tuples can be counted in the following way: first we fix a k-tuple K, then select s r-tuples consisting entirely of vertices of K, and m - s r-tuples from among the remaining $\binom{n}{r}$ - s. Thus

$$\Sigma \quad b(H^{(r)}) \leq \binom{n}{k} \binom{\binom{k}{r}}{s} \binom{\binom{n}{r} - s}{m - s}$$

and hence

$$(2) \quad \Sigma \quad b(H^{(r)}) \div \binom{\binom{n}{r}}{m} \leq c_1 n^k \frac{\binom{\binom{n}{r} - s}{m - s}}{\binom{\binom{n}{r}}{m}}$$

$$\leq c_1 n^k \left(\frac{m}{\binom{n}{r}}\right)^s$$

where c_1 is a constant depending on k,r, and s. The last inequality is a special case of the following:

If $B \leq C \leq A$ then $(C-i)/(A-i)$ is a decreasing function of i and so

$$\frac{\binom{A - B}{C - B}}{\binom{A}{C}} = \frac{C(C-1)\ldots(C-B+1)}{A(A-1)\ldots(A-B+1)} \leq (C/A)^B.$$

Inequality (2) will imply (1), provided

$$c_1 n^k \left(\frac{m}{\binom{n}{r}}\right)^s < c_2 m$$

i.e., $m^{s-1} < c_3 n^{rs-k}$; thus we may fix $m = c_4 \, n^{(rs-k)/(s-1)}$ thereby proving the theorem.

5. The order of magnitude of $f^{(3)}(n;k,k-2)$.

By our theorem, $f^{(3)}(n;k,k-2) > cn^2$. We prove that (constant) x n^2 triples suffice to ensure the existence of a $G^{(3)}(k,k-2)$. Let

$$G^{(3)} = G^{(3)}\left(n,\frac{1}{3}\left(n\left[\frac{k-2}{k-1}\cdot(n-1)\right] + 1\right)\right).$$

Then some vertex x has the property that the pairs of vertices which together with x constitute triples of the 3-graph form a

$$G^{(2)}\left(n-1, \left[\frac{k-2}{k-1}(n-1)\right] + 1\right)$$

on the vertex set of our $G^{(3)}$ with x omitted. By a result quoted in Section 2, such a graph must contain a $G^{(2)}(k-1,k-2)$, hence $G^{(3)}$ contains a $G^{(3)}(k,k-2)$.

We conjecture that $\lim_{n\to\infty} n^{-2}f^{(3)}(n;k,k-2)$ exists, but have succeeded [4] in proving this only for $k = 4$.

Many interesting new problems arise if we also consider the structure of the graphs $G^{(r)}(k;s)$ and we have some very preliminary results; but many unsolved problems remain even for $r = 2$.

References

[1] C. Benson, Minimal regular graphs of girth eight and twelve, Canad. J. Math., 18 (1966), 1091-1094.

[2] C. Berge, Graphes et Hypergraphes, Dunod, Paris, 1971.

[3] W. G. Brown, On graphs that do not contain a Thomsen graph, Canad. Math. Bull., 9 (1966), 281-285.

[4] W. G. Brown, P. Erdös, V. T. Sós, On the existence of triangulated spheres in 3-graphs and related problems, Studio Sci. Math. Hungar., to appear.

[5] P. Erdös, Extremal problems in graph theory, Theory of Graphs and Its Applications, (M. Fiedler, ed.) Academic Press, New York, 1964, 29-36.

[6] P. Erdös, Graph theory and probability I, Canad. J. Math., 11 (1959), 34-38.

[7] P. Erdös, A. Rényi, V. T. Sós, On a problem of graph theory, Studio Sci. Math. Hungar., 1 (1966), 215-235.

[8] P. Erdös, M. Simonovits, A limit theorem in graph theory, Studio Sci. Math. Hungar., 1 (1966), 51-57.

[9] W. Mantel (proposer), W. A. Wythoff (author of one solution), Vraagstuk XXVIII, Wiskundige Opgaven met de Oplossingen, 10 (1907), 60.

[10] R. W. Singleton, On minimal graphs of maximum even girth, J. Combinatorial Theory, 1 (1966), 306-332.

[11] P. Turán, Egy gráfelméleki Szélsöérték-feladatról, Mat. Fiz. Lapok, 48 (1941), 436-452.

NEW DIRECTIONS IN HAMILTONIAN GRAPH THEORY

Václav Chvátal

1. Introduction.

Ever since Sir William Hamilton's venture into the world of business (see [24, p. 5] for more details), many brilliant minds were occupied with the question "Which graphs are hamiltonian?". In this paper, we collect some more recent developments related to this question, show the links among them and indicate which directions of research on this subject seem to be most promising. We do not intend to give a complete survey of all related results. For the material not covered here, the reader is referred to the expository paper [32] by Nash-Williams and to Section 1.4 of Grünbaum [23].

Our notation and terminology follows Harary [25] with only minor modifications; notation not explained here can be found there. The order of a graph $G = (V,E)$ is the number of points (vertices); the size of G is the number of lines (edges). As in [25], the order of G will be denoted by p and its size by q. If $H = (V_o, E_o)$ is a subgraph of $G = (V,E)$ and if T is a subset of $V - V_o$ then we denote by $q(H,T)$ the number of lines of G having one endpoint in V_o and the other in T. Similarly, $q(T)$ is the size and $k(T)$ the number of components of the subgraph $\langle T \rangle$ of G induced by T. If F is a spanning subgraph of G, we write $F < G$. The set of all graphs of order p, partially ordered by the relation <, will be denoted by $\mathcal{G}_p$. A subset S of a partially ordered set P is an <u>upper order ideal</u> if $x \in S$, $x < y$ implies $y \in S$. Similarly, S is a <u>lower order ideal</u> if $x \in S$, $y < x$ implies $y \in S$.

The question "Which graphs are hamiltonian?" does not make much sense unless we specify what form of an answer we want. A list of hamiltonian graphs? It would have to be an infinite one. An algorithm for finding a hamiltonian cycle? We have one -- the obvious trial and error method -- but it is too slow; how efficient should it be to satisfy you? Perhaps the most sensible thing to want is a <u>good characterization</u> of nonhamiltonian graphs. Here the words "good characterization" are to be understood in their technical sense as introduced by Edmonds [17]. To illustrate this quite basic and stimulating concept, we shall paraphrase Edmonds' tale of the "absolute supervisor".

A rich and powerful king, driven by his aristocratic whims, desires to know whether a certain graph (with an awful lot of points and lines) has a 1-factor. He hires a team of mathematicians and they set to work. If they eventually find a 1-factor, they can just show it to the king; it won't take him long to check that he wasn't cheated. However, what happens if there is no 1-factor in the graph? Will the king take their word for it? Won't he suspect this bunch of intellectuals of doing nothing, only pretending to work? Spending their time at wild orgies? No. If G has no 1-factor, then there is a set S of points such that G - S has more than |S| odd components; this is Tutte's theorem [41]. Therefore the king does not need to trust his mathematicians. If they cannot find a 1-factor, they had better find that set S. Needless to say, they may have a hard time finding it; however, it will be easy -- for the king -- to check that they did their job. And this is exactly what we mean when we say that Tutte's theorem provides a good characterization of graphs without 1-factors.

2. Status Quo.

Although the problem of the existence of a 1-factor is, from the formal viewpoint, completely analogous to the problem of the existence of a hamiltonian cycle (both a 1-factor and a hamiltonian cycle are spanning subgraphs), we have no good characterization of nonhamiltonian graphs. Nevertheless, there are several theorems giving necessary *or* sufficient conditions for the existence of a hamiltonian cycle; a few of these are listed below.

Theorem 1. (Tutte [43]). Every 4-connected planar graph is hamiltonian.

Theorem 2. (Chvátal [9]). If the degrees $d_1 \leq d_2 \leq \ldots \leq d_p$ of the points of G satisfy

(1) $d_m \leq m < p/2$ implies $d_{p-m} \geq p - m$,

then G is hamiltonian.

Theorem 3. (Fleischner [22]). If G is 2-connected, then G^2 is hamiltonian.

Theorem 4. (Chvátal and Erdös [13]). If the point-independence number $\beta_o(G)$ does not exceed its connectivity $\kappa(G)$, then G is hamiltonian.

Theorem 5. (Nash-Williams and Bondy [31]). If the point-independence number $\beta_o(G)$ does not exceed the minimum degree $\delta(G)$ and if $\delta(G) \geq (p+2)/3$, then G is hamiltonian.

Theorem 6. (Chvátal [12]). If $G = (V,E)$ is hamiltonian, then there is no partition $V = R \cup S \cup T$ with

$$(2) \qquad T \neq V \text{ and } |S| + \sum_H q(H,T)/2 < k(T)$$

where the summation is extended over all components H of $\langle R \rangle$.

Theorem 1 strengthens a previous theorem of Whitney [45] who established the existence of a hamiltonian cycle in every 4-connected maximal planar graph. However, Tutte [40] constructed a cubic 3-connected planar nonhamiltonian graph (of order 46), thus showing that Theorem 1 is in a sense best possible.

Let us also note the peculiar form of Theorem 1: its conclusion asserts that G belongs to a certain upper order ideal (namely the set of hamiltonian graphs in G_p, while the hypotheses specify that G belongs to the intersection of an upper order ideal (4-connected graphs) with a lower order ideal (planar graphs). On the other hand, Theorems 2, 4 and 5 assert that certain upper order ideals in $\mathcal{G}_p$. consist of hamiltonian graphs only. What other non-trivial theorems of this form could one prove? Duke [16] showed that the following analog of Theorem 1 for graphs of higher genus γ holds: Let the function $c(\gamma)$ satisfy the inequalities

$$4 \leq c(1) \leq 6 \text{ and}$$

$$[\tfrac{1}{2}(5 + (16\gamma + 1)^{1/2})] \leq c(\gamma) \leq [4 + (6\gamma + 3)^{1/2}] \quad \text{for } \gamma > 1.$$

Then every $c(\gamma)$-connected graph of genus γ is hamiltonian. It would be interesting to find out whether the lower order ideal of planar graphs in Theorem 1 can be replaced by a larger one. Perhaps the following holds.

<u>Conjecture 1</u>. If G is 4-connected and every set of n points in G induces a graph of size $\geq 3n - 6$, then G is hamiltonian.

Finally, it would be interesting to find a simple class of graphs which all have 4-connected planar spanning subgraphs (and therefore are hamiltonian).

Next, we turn our attention to Theorem 2. Nash-Williams [33] suggested that a degree sequence

(3) $$d_1 \leq d_2 \leq \cdots \leq d_p$$

be called <u>forcibly hamiltonian</u> if every graph with this degree sequence is necessarily hamiltonian; <u>forcibly non-hamiltonian</u> if every graph with this degree sequence is necessarily non-hamiltonian; and finally, <u>optionally hamiltonian</u> if it is neither forcibly hamiltonian nor forcibly non-hamiltonian. If $S = (d_1, d_2, \ldots, d_p)$, $S^* = (d_1^*, d_2^*, \ldots, d_p^*)$ are non-decreasing degree sequences with $d_i \leq d_i^*$ for each $i = 1,2,\ldots,p$, then we write $S < S^*$ and say that S^* <u>dominates</u> S. Then the set $\mathcal{S}_p$ of all nondecreasing degree sequences of length p forms a partially ordered set; the canonical mapping $f : \mathcal{G}_p \rightarrow \mathcal{T}_p$ is order-preserving, i.e., $G < H$ implies $f(G) < f(H)$. Since the hamiltonian graphs form an upper order ideal in $\mathcal{G}_p$, one might expect the forcibly hamiltonian sequences to form an upper order ideal in $\mathcal{S}_p$. However, this is not the case. Consider the degree sequences

$$S_1 = (2,2,2,2,2),$$
$$S_2 = (2,2,2,3,3),$$
$$S_3 = (2,2,2,4,4).$$

One has $S_1 < S_2 < S_3$ but S_1 is forcibly hamiltonian, S_2 is optionally hamiltonian and S_3 is forcibly nonhamiltonian.

In the light of this observation, it is natural to ask what is the largest upper order ideal contained in the set of forcibly hamiltonian sequences of length p. The answer is provided by Theorem 2. If a degree sequence (3) fails to satisfy (1) then it is dominated by the sequence

$$\underbrace{m,m,\ldots,m}_{m \text{ times}};\quad \underbrace{p-m-1,p-m-1,\ldots,p-m-1}_{p-2m \text{ times}};\quad \underbrace{p-1,\ldots,p-1}_{m \text{ times}}$$

which is forcibly nonhamiltonian; indeed, the only graph with this degree sequence is $K_m + (\overline{K}_m \cup K_{p-2m})$. On the other hand, it is easy to see that degree sequences satisfying (1) form an upper order ideal in $\mathcal{S}_p$. Therefore (1) characterizes the largest upper order ideal contained in the set of all forcibly hamiltonian degree sequences of length p. The graphic degree sequences (which belong to graphs) have been characterized by Erdös and Gallai [21] among others, see [25, Chapter 6]. Previous theorems giving sufficient conditions for a graphic sequence to be forcibly hamiltonian are due to

Dirac [15], Pósa [37] and Bondy [3]. Pósa's theorem is stronger than Dirac's and Bondy's theorem is stronger than Pósa's. All three, however, determine certain upper order ideals in S_p and assert that these ideals consist of forcibly hamiltonian sequences only. Obviously, these ideals are all contained in the largest one and so Theorem 2 is stronger than the three previous ones.

There is a striking formal similarity between Theorem 2 and a recent theorem of Erdös [20]. Erdös' theorem asserts that every graph G containing no K_m is degree-majorized by an (m-1) colorable graph H, i.e., there is a one-to-one correspondence $f : V(G) \to V(H)$ with $d_G(u) \leq d_H(f(u))$ for all $u \in V(G)$. Similarly, Theorem 2 asserts that every graph G containing no hamiltonian cycle is degree-majorized by the graph $H = K_m + (\overline{K}_m \cup K_{p-2m})$ with a suitable m, $0 < m < p/2$. Erdös' theorem readily yields the celebrated Turán's formula [39] for $ex(p,K_m)$ and a characterization of the corresponding extremal graphs. Similarly, as observed by Bondy [4], Theorem 2 yields at once Ore's formula [35]:

$$ex(p,C_p) = 1 + \binom{p-1}{2}.$$

Indeed, the size of $K_m+(\overline{K}_m \cup K_{p-2m})$ is $f(p,m) = \binom{p-m}{2} + m^2$. By Theorem 2, the maximal possible size of a nonhamiltonian graph is

$$\max_{0<m<p/2} f(p,m) = f(p,1) = 1 + \binom{p-1}{2}.$$

Besides, we have $f(p,m) < f(p,1)$ whenever $m > 1$ and $(p,m) \neq (3,2)$. Therefore the only nonhamiltonian graphs of size $1 + \binom{p-1}{2}$ are $K_1 + (\overline{K}_1 \cup K_{p-2})$ and $K_2 + (\overline{K}_2 \cup K_1) + K_2 = \overline{K}_3 + K_2$.

Let us also note that the deletion of m points (those corresponding to K_m) from $H = K_m + (\overline{K}_m \cup K_{p-2m})$ results in the graph $\overline{K}_m \cup K_{p-2m}$ having m+1 components. This could never happen if H were hamiltonian. Indeed, every hamiltonian graph G satisfies

$$(4) \qquad S \neq \emptyset \text{ implies } k(G-S) \leq |S|.$$

In Section 4, graphs satisfying (4) are called <u>1-tough</u>. Hence Theorem 2 asserts that if a degree sequence is not forcibly hamiltonian then it is dominated by the degree sequence of a graph which is not 1-tough. One might suspect that each degree sequence which is not forcibly hamiltonian is the degree sequence of a graph which is not 1-tough.

However, this is not true. The only graph with degree sequence (6, 4, 4, 4, 2, 2, 2) is the graph G in Figure 2; G is 1-tough and non-hamiltonian. Some of the forcibly hamiltonian sequences which fall out of the upper order ideal (1) were characterized by Nash-Williams [33].

Theorem 3 has an interesting history. Some years ago, Sekanina [38] proved that G^3 is hamiltonian whenever G is connected. Plummer and Nash-Williams [30], independently of each other, were led to conjecture that G^2 is hamiltonian whenever G is 2-connected. Their conjecture remained open for about three years, until in January 1971, Herbert Fleischner [ ] proved its validity. Even more recently, Hobbs with Nash-Williams and Jung found independently the same simple way of showing that Theorem 3 implies its own strengthening: if G is 2-connected, then G^2 is hamiltonian-connected, i.e., each pair of its points is joined by a hamiltonian path. Since the squares of 2-connected graphs are rather rare animals, it would be desirable to specify which of their properties force them to be hamiltonian; we shall pursue this idea further in Section 4.

Theorem 4 is obviously best possible: the complete bipartite graph $K_{m,m+1}$ is m-connected, nonhamiltonian and its point-independence number is m+1. However, it would be silly to believe that the graphs $K_{m,m+1}$ are the only nonhamiltonian graphs with $\beta_o = k+1$. The Petersen graph is another example of a nonhamiltonian graph with $\kappa = 3$, $\beta_o = 4$. If $\delta \geq (p+2)/3$, then Theorem 5 is stronger than Theorem 4 as $\kappa \leq \delta$.

The necessary condition given by Theorem 6 is far from sufficient. Nevertheless, it approximates the desired characterization quite well in at least two respects. Indeed, if G admits no partition (2), then G is 1-tough and has a 2-factor. To see that G must be 1-tough, it suffices to realize that (2) is satisfied by $R = \emptyset$, $T = V-S$ whenever (4) is violated. If G has no 2-factor, then there is a partition $V = R \cup S \cup T$ with

$$(5) \qquad |S| = \sum_H q(H,T)/2 < |T| - q(T);$$

this is a special case of Tutte's factor theorem [42]. Since G admits no partition (2) and one has $|T| - q(T) \leq k(T)$, we must necessarily have $T = V$. But then (5) reads $q(T) < |T|$, i.e., $q < p$, and G cannot be 1-tough (not even 2-connected) -- a contradiction. However, the condition given by Theorem 6 is stronger than just 1-toughness and the existence of a 2-factor combined together. The graph in Figure 1 is 1-tough and has a 2-factor but admits a partition (2) as indicated.

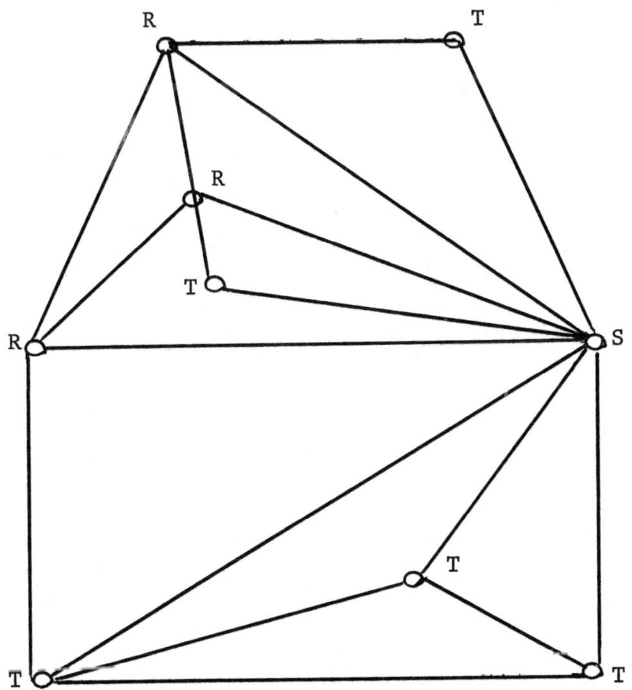

Figure 1

3. Cyclability.

We have observed that every hamiltonian graph is 1-tough and therefore 2-connected. However, the 2-connectedness of G is far from implying that G is hamiltonian. Nevertheless, by the classical theorem of Menger [29], a graph is 2-connected if and only if every two points of G lies on a common cycle. Thus, we are led to say that a graph is n-cyclable if and only if every n of its points lie on a common cycle. By Menger's theorem, "2-cyclable" is the same as "2-connected"; on the other hand, "p-cyclable" is the same as "hamiltonian". The cyclability $\xi(G)$ of G is the largest n such that G is n-cyclable. The following link between connectivity and cyclability has been established by Dirac [14].

Theorem 7. $\kappa \geq 2$ implies $\xi \geq \kappa$.

Watkins and Mesner [44] found a good characterization of graphs which are not 3-cyclable.

Theorem 8. A graph $G = (V,E)$ is not 3-cyclable if and only if one of the following conditions holds:

(i) $\kappa \leq 1$ (then G is not even 2-cyclable),

(ii) there is a set $S \subset V$ with $|S| = 2$, $k(G-S) \geq 3$,

(iii) $V = V_o \cup V_1 \cup V_2 \cup V_3$ where each line of G has both endpoints in one of the sets V_o, V_1, V_2, V_3 and

$$V_1 \cap V_2 = V_1 \cap V_3 = V_2 \cap V_3 = \{v\} \subset V_o$$

$$|V_i \cap V_o| = 2,\ V_i - V_o \neq \emptyset\ (i = 1,2,3),$$

(iv) $V = W_1 \cup W_2 \cup V_1 \cup V_2 \cup V_3$ where each line of G has both endpoints in one of the sets W_1, W_2, V_1, V_2, V_3 and

$$W_1 \cap W_2 = \emptyset,\ V_1 \cap V_2 = V_1 \cap V_3 = V_2 \cap V_3 = \emptyset$$

$$|V_i \cap W_j| = 1,\ |V_i| \geq 3\ (i=1,2,3;\ j=1,2).$$

Besides, Watkins and Mesner [44] also characterized graphs with $\xi = \kappa \geq 3$.

Theorem 9. $\xi(G) = \kappa(G) \geq 3$ if and only if there is a set $S \subset V$ with $|S| = \kappa$, $\kappa(G-S) = k+1$.

These two theorems of Watkins and Mesner contribute greatly to our understanding of the structure of nonhamiltonian graphs. It would be quite desirable to extend their results by characterizing graphs which are not 4-cyclable. This is perhaps not extremely difficult (actually, one would have to consider only the graphs with $\kappa = 2$ and $\xi = 3$) but no such characterization has been given yet.

Conjecture 2. If the addition of any new line to a graph G increases its cyclability, then it results in a hamiltonian graph. (In other words, the maximal non-n-cyclable graphs are always maximal nonhamiltonian).

Conjecture 3. If G is nonhamiltonian, then it is degree-majorized by a graph H containing a set S of points with $|S| \leq \xi(G)$, $k(G-S) = |S| + 1$.

Conjecture 2 would introduce an orderly system to the set of maximal nonhamiltonian graphs; Conjecture 3 would provide a natural refinement of Theorem 2.

One might hope that the approach through cyclability, characterizing successively the graphs which are not n-cyclable for n = 4,5,6,..., would eventually lead to a characterization of nonhamiltonian graphs. However, there are reasons to be pessimistic. There exist graphs of cyclability p-1; there are many of them and they seem to be quite difficult to characterize. These graphs are now called hypohamiltonian and were studied by Sousselier (see [1], [2]) more than forty years ago. Sousselier found that the smallest hypohamiltonian graph is the Petersen graph and also constructed an infinite sequence of other hypohamiltonian graphs [26]. Since that time, several other authors [26],[27],[4],[10] have been working on this subject.

Theorem 10. (Chvátal [10]). There exist hypohamiltonian graphs of every order ≥ 42. Further, the number of non-isomorphic hypohamiltonian graphs of order p tends to infinity with p.

The hypohamiltonian graphs present several amusing problems. Is there a planar hypohamiltonian graph? If so, can such a graph be cubic? Is there a graph G such that no hypohamiltonian graph has a subgraph isomorphic to G? Herz, Duby and Vigué [26] conjecture that every hypohamiltonian graph has girth ≥ 5. What is the maximal possible size of a hypohamiltonian graph of order p?

4. Toughness.

We will say that a graph G is t-tough if

$$k(G-S) > 1 \text{ implies } |S| \geq t \cdot k(G-S).$$

If G is not complete then there is a largest t such that G is t-tough. This value of t will be called the toughness of G and denoted by $t(G)$; we will also set $t(K_m) = +\infty$. Adopting the convention $\min \emptyset = +\infty$ we can write

$$t(G) = \min \frac{|S|}{k(G-S)}$$

where S runs through all the point-cutsets of G. It is easy to see that

(6) $\qquad G < H$ implies $t(G) \leq t(H)$.

Toughness is a non-decreasing invariant whose values range from zero to infinity; $t(G) = 0$ if and only if G is disconnected, $t(G) = +\infty$ if and only if G is complete. Below we list a few interesting properties of toughness.

(7) $\quad t \geq \frac{\kappa}{\beta_o}$

(8) $\quad t \leq \frac{\kappa}{2} \quad (\kappa \leq p-2)$

$\quad t \leq \frac{p-\beta_o}{\beta_o} \quad (\beta_o \geq 2)$

$\quad t(K_{m,n}) = \frac{m}{n} \ (m \leq n)$

$\quad t(Q_n) = 1 \quad (n \geq 2)$

(9) $\quad t(C_p^m) = m \quad (2m + 1 < p)$

$\quad t(K_m \times K_n) = \frac{m+n}{2} - 1 \quad (m,n \geq 2)$

(10) $t(G^2) \geq \kappa(G)$.

The main importance of toughness lies in its relationship to hamiltonian circuits. It follows readily from (9) with $m = 1$ and from (6) that

(11) G is hamiltonian implies $t(G) \geq 1$.

Unfortunately, the converse of (11) holds only for graphs with at most six points; the nonhamiltonian graph in Figure 2 is 1-tough.

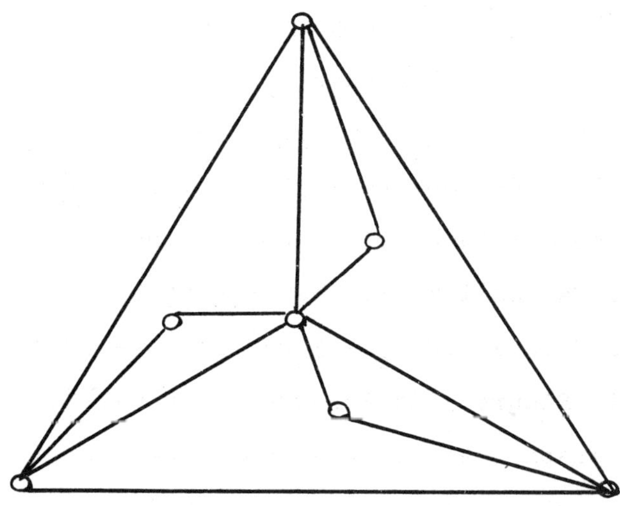

Figure 2

Nevertheless, we have seen that Theorem 2 can be interpreted as a weak converse of (11). Similarly, Theorem 4 provides another weak converse of (11): if the lower bound for t given by (7) is at least 1, then G is hamiltonian.

Conjecture 4. There is a t_o such that every t_o-tough graph is hamiltonian.

A proof of this conjecture with $t_o = 2$ would yield a generalization of Theorem 3; indeed, by (10), the square of a 2-connected graph is 2-tough. Incidentally, the squares of 2-connected graphs enjoy another interesting property (apart from being 2-tough): the neighborhood of each point in G^2 induces a connected subgraph (this is true for all graphs G). Perhaps this property and 2-toughness imply the existence of a hamiltonian circuit.

As far as I know, the largest t_o for which Conjecture 4 fails is $t_o = 3/2$; the non-hamiltonian graph in Figure 3 is 3/2-tough. However, every planar graph that is more than 3/2-tough is necessarily hamiltonian; this follows directly from (8) and Theorem 1. For more information on toughness, the reader is referred to [11].

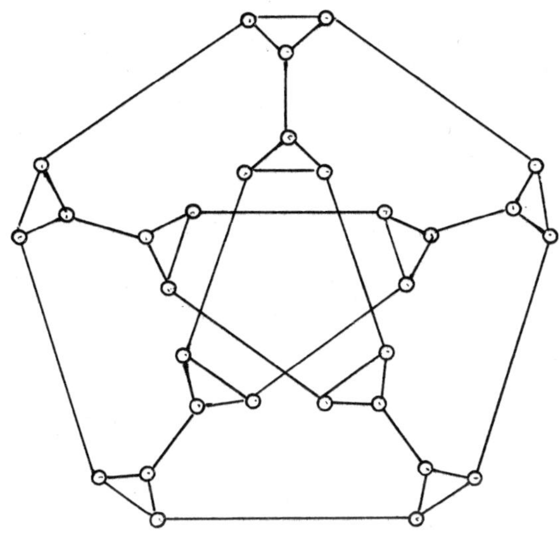

Figure 3

5. Pancyclic Graphs.

There are many variations on the hamiltonian problem. One may investigate the graphs having a hamiltonian path, the hamiltonian-connected graphs [36], the n-hamiltonian graphs [8] (which remain hamiltonian after the deletion of any $m \leq n$ points), and so on. Nevertheless, these variations turn out to be equivalent to the original problem, in the sense that a good characterization of hamiltonian graphs could be converted into a good characterization of hamiltonian-connected graphs, etc. -- and vice versa.

However, a genuinely non-equivalent variation has been recently introduced by Bondy's concept of "pancyclic graphs". A graph is called _pancyclic_ if it has cycles of every length n, $3 \leq n \leq p$. By definition, every pancyclic graph is hamiltonian. A condition under which the converse holds is given by the next result, in which d(v) is the degree of v.

Theorem 11. (Bondy [5]). Every hamiltonian graph of order p and size at least $p^2/4$ is either pancyclic or else it is the complete bipartite graph $K_{p/2,p/2}$.

Ore [34] proved that if

$$\{u,v\} \notin E \text{ implies } d(u) + d(v) \geq p,$$

then G is hamiltonian. Theorem 11 yields at once the following strengthening of Ore's result: every graph satisfying (12) is either pancyclic or else it is the complete bipartite graph $K_{p/2,p/2}$.

This observation illustrates a peculiar feature of pancyclic graphs which led Bondy [6] to the following idea.

Metaconjecture. Almost any non-trivial condition on a graph which implies that the graph is hamiltonian also implies that the graph is pancyclic. (There may be a simple family of exceptional graphs.)

For example, a special case of Theorem 4 has been strengthened by Bondy [6] as follows: Every 2-connected graph G with $\beta_o(G) \leq 2$ is either pancyclic or else G is one of the cycles C_4 or C_5. Any such generalization of Theorem 4 with $\beta_o \leq n$ would have to take account of at least two families of non-pancyclic graphs: the complete bipartite graphs $K_{m,n}$ and the graphs G_n consisting of cycles $u_1, u_2, \ldots, u_{2n+2}$ with additional lines $\{u_i, u_{i+2k}\}$, $1 < k < n$, $1 \leq i \leq 2n+2$.

One can measure the goodness of each existence theorem for hamiltonian cycles by the size and structure of the "exceptional" family of non-pancyclic graphs. For instance, Malkevitch [28] has constructed an infinite class of 4-connected planar non-pancyclic graphs; this indicates the depth of Theorem 1.

Among the conjectures suggested by Bondy's meta-conjecture, the following two are still unsettled.

Conjecture 5. If the degree sequence of G satisfies (1), then G is pancyclic or bipartite.

Conjecture 6. If G is 2-connected then G^2 is pancyclic.

For more information on pancyclic graphs, the reader is referred to [5],[6] and [7].

6. Weakly Hamiltonian Graphs.

The hamiltonian problem is essentially an integer linear programming problem. Indeed, the characteristic function of a hamiltonian cycle in $G = (V,E)$ is an integer-valued function f satisfying the following constraints in which $f \cdot S(u)$ is the sum of $f(x)$ over all lines x incident with u and $f \cdot [T]$ is the sum of $f(x)$ over all lines x having both endpoints in T.

$$(12) \quad \begin{cases} f(x) \geq 0 & (x \in X) \\ f \cdot S(u) \leq 2 & (u \in V) \\ f \cdot [T] \leq |T| - 1 & (T \subset V,\ \emptyset \neq T \neq V) \end{cases}$$

Conversely, every integer-valued function $f : E \to [0,\infty)$ satisfying (12) and

$$(13) \qquad f \cdot [V] = p$$

is the characteristic function of a hamiltonian circuit in G. The inequalities (12) determine a polyhedron P in the q-dimensional Euclidean space R^E. To decide whether G is hamiltonian, we have to maximize $f \cdot [V]$ over the set H of all integral (lattice) points $\{f(x) : x \varepsilon E\}$ included in the polyhedron P. Obviously, the set H is finite; its convex hull is a polyhedron E(P), called the _Edmonds polyhedron_ of P. It is not difficult to see that the maximum of $f \cdot [V]$ over _all_ (not necessarily integral) points in E(P) equals the maximum of $f \cdot [V]$ over all _integral_ points in P. In other words, the constraints (12) plus the integrality of f imply other linear constraints on f; the knowledge of the complete set of all such constraints would enable us to convert the _integer_ linear programming problem into an _ordinary_ (continuous) linear programming problem. The great advantage of such a conversion lies in the applicability of the duality theorem of linear programming: if we knew the inequalities determining E(P), then the duality theorem would yield at once a good characterization of nonhamiltonian graphs. In the context of factors with prescribed degrees, this approach has been carried out successfully by Edmonds [18],[19]. Its application in the present context was suggested by Nash-Williams.

In [12], it is proved that every integer-valued function satisfying (12) must also satisfy the "comb inequality"

$$(14) \qquad f \cdot K \leq r(K).$$

There $K = [W_o] \cup [W_1] \cup \ldots \cup [W_n]$ where each $W_i (i = 0,1,\ldots,n)$ is a non-empty proper subset of V and $|W_i \cap W_o| = 1$ for each $i = 1,2,\ldots,n$. The number $f \cdot K$ is the sum of $f(x)$ over all lines x having both endpoints in the same W_i while $r(K)$, the rank of the comb K, is defined to be

$$|W_o| + \sum_{i=1}^{n} (|W_i| - 1) - \{\tfrac{n}{2}\}.$$

A _function_ $f : E \to [0,+\infty)$ is said to be _weakly hamiltonian_ if it satisfies the constraints (12) and (14); a _graph_ G is said to be _weakly hamiltonian_ if it admits a weakly hamiltonian function satisfying (13). Obviously, every hamiltonian graph is weakly hamiltonian. The duality theorem of linear programming yields a characterization of weakly hamiltonian graphs which can be seen as a necessary condition for a graph G to be hamiltonian. This condition is stronger but less elegant than Theorem 6. The following "one-two-three theorem" gives a rough idea as to how closely weakly hamiltonian graphs approximate the hamiltonian ones.

Theorem 12. (Chvátal [12]). Every weakly hamiltonian graph is 1-tough, has a 2-factor and is

3-cyclable.

Consequently, every weakly hamiltonian graph with less than 10 points is hamiltonian. This bound is best possible since the Petersen graph is weakly hamiltonian; the corresponding weakly hamiltonian function is defined by f(x) = 2/3. For more information on weakly hamiltonian graphs, the reader is referred to [12].

The challenge is to find new linear inequalities implied by (12) and the integrality of f and to obtain thus -- via the duality theorem of linear programming -- stronger necessary condition for the existence of a hamiltonian cycle in G.

ADDED IN PROOF.

1. The following bipartite graph due to B. Grünbaum provides a counterexample to Conjecture 1. Its 16 vertices are a,b,...f and A,B,...,J; its edges are given by the matrix:

	A	B	C	D	E	F	G	H	J
a	1	0	0	1	1	1	1	0	1
b	1	1	0	0	1	1	1	1	0
c	1	1	1	0	0	1	0	1	1
d	1	1	1	1	0	0	1	0	1
e	0	1	1	1	1	0	1	1	0
f	0	0	1	1	1	1	0	1	1

2. Trying to extend our Theorem 2, Las Vergnas [46] proved the following:

Let G be a graph with points $u_1, u_2, \ldots, u_p$ where the degree sequence $d(u_1), d(u_2), \ldots, d(u_p)$ is not necessarily monotone; let s be an integer with $0 \leq s \leq p-2$. Let us assume that (1) $1 \leq i < j \leq p$, (2) $j \geq p - s - i$, (3) $d(u_i) \leq i + s$, (4) $d(u_j) \leq j - 1 + s$, and (5) u_i not adjacent to u_j together imply $d(u_i) + d(u_j) \geq p + s$. Then every set of at most s lines of G that form pairwise disjoint paths is contained in a hamiltonian cycle of G.

3. In the spirit of Bondy's Metaconjecture, I. Zarins conjectured the existence of (the smallest) p(k) such that every hamiltonian graph with $\beta_0(G) \leq k$ and at least p(k) points is necessarily pancyclic. Bondy's result yields p(2) = 6; Zarins proved $p(3) \leq 416$. During the Hypergraph Conference at Ohio State University (August, 1972), P. Erdös proved Zarins' conjecture. Actually, he proved that

$$c_1 k^3 \leq p(k) \leq c_2 k^9.$$

4. J. A. Bondy found a simple and elegant proof of Conjecture 5 with p odd. In this case, condition (1) implies that more than p/2 vertices of G have

degrees exceeding p/2. Therefore, a hamiltonian cycle of G contains two consecutive points u,v with d(u) + d(v) > p. A fortiori (see the proof of the main theorem in [5]), G is pancyclic.

5. A systematic way of generating new linear inequalities implicit in (12) was developed by the author (and illustrated on the Petersen graph) in a forthcoming paper [47].

REFERENCES

[1] C. Berge, Problèmes plaisants et délectables, Rev. Francaise Informat. Recherche Opérationelle 29 (1963), 405-406.

[2] C. Berge, Graphes et Hypergraphes, Dunod, Paris 1970, 213-217.

[3] J. A. Bondy, Properties of graphs with constraints on degrees, Studia Sci. Math. Hungar. 4 (1969), 473-475.

[4] J. A. Bondy, Variations on the Hamiltonian Theme, Canad. Math. Bull. 15 (1972), 57-62.

[5] J. A. Bondy, Pancyclic graphs I., J. Combinatorial Theory 11B (1971), 80-84.

[6] J. A. Bondy, Pancyclic graphs II., J. Combinatorial Theory, to appear.

[7] J. A. Bondy, Pancyclic graphs III., to appear.

[8] G. Chartrand, S. F. Kapoor and D. R. Lick, n-hamiltonian graphs, J. Combinatorial Theory 9 (1970), 308-312.

[9] V. Chvátal, On Hamilton's ideals, J. Combinatorial Theory 12 (1972), 163-168.

[10] V. Chvátal, Flip-flops in hypohamiltonian graphs, Canad. Math. Bull., to appear.

[11] V. Chvátal, Tough graphs and hamiltonian circuits, Discrete Math., to appear.

[12] V. Chvátal, Edmonds polyhedra and weakly hamiltonian graphs, J. Math. Programming, to appear.

[13] V. Chvátal and P. Erdös, A note on hamiltonian circuits, Discrete Math. 2 (1972), 111-113.
[14] G. A. Dirac, In abstrakten Graphen vorhandene vollständige 4-Graphen und ihre Unterteilungen, Math. Nachr. 22 (1960), 61-85.
[15] G. A. Dirac, Some theorems on abstract graphs, Proc. London Math. Soc. 2 (1952), 69-81.
[16] R. A. Duke, On the genus and connectivity of hamiltonian graphs, Discrete Math. 2 (1972), 199-206.
[17] J. Edmonds, Minimum partition of a matroid into independent sets, J. Res. Nat. Bur. Standards 69B (1965), 67-72.
[18] J. Edmonds, Maximum matching and a polyhedron with 0,1-vertices, J. Res. Nat. Bur. Standards 69B (1965), 125-130.
[19] J. Edmonds, Paths, trees and flowers, Canad. J. Math. 17 (1965), 499-467.
[20] P. Erdös,
[21] P. Erdös and T. Gallai, Graphs with prescribed degrees of vertices (Hungarian), Mat. Lapok 11 (1960), 264-274.
[22] H. Fleischner, The square of every two-connected graph is hamiltonian, J. Combinatorial Theory, to appear.
[23] B. Grünbaum, Polytopes, graphs and complexes, Bull. Amer. Math. Soc. 76 (1970), 1131-1201.
[24] F. Harary, A Seminar on Graph Theory, Holt, New York, 1967.
[25] F. Harary, Graph Theory, Addison-Wesley, Reading, 1969.
[26] J. C. Herz, J. J. Duby, F. Vigué, Recherche systématique des graphes hypohamiltoniens, Théorie des Graphes (P. Rosenstiehl, ed.), Dunod, Paris, 1967, 153-160.
[27] W. F. Lindgren, An infinite class of hypohamiltonian graphs, Amer. Math. Monthly 74 (1967), 1087-1088.
[28] J. Malkevitch, On the lengths of cycles in planar graphs, Recent Trends in Graph Theory (M. Capobianco et al., eds.), Springer-Verlag, 1971.
[29] K. Menger, Zur allgemeinen Kurventheorie, Fund. Math. 10 (1927), 96-115.
[30] C. St. J.A. Nash-Williams, Problem 48, Theory of Graphs (P. Erdös and G. Katona, eds.), Akadémiai Kiadó, Budapest, 1968.

[31] C. St. J.A. Nash-Williams, Edge-disjoint hamiltonian circuits in graphs with vertices of large valency, Studies in Pure Mathematics, (L. Mirsky, ed.), Academic Press, London, 1971.

[32] C. St. J.A. Nash-Williams, Hamiltonian arcs and circuits, Recent Trends in Graph Theory, (M. Capobianco et al., eds.), Springer-Verlag, Berlin, 1971.

[33] C. St. J.A. Nash-Williams, Unexplored and semi-explored territories in graph theory, this volume.

[34] O. Ore, Note on hamiltonian circuits, Amer. Math. Monthly 67 (1960), 55.

[35] O. Ore, Arc coverings of graphs, Ann. Mat. Pura Appl. 55 (1961), 315-322.

[36] O. Ore, Hamilton-connected graphs, J. Math. Pures Appl. 42 (1963), 21-27.

[37] L. Pósa, A theorem concerning hamilton lines, Magyar Tud. Akad. Mat. Kutato Int. Közl. 7 (1962), 225-226.

[38] M. Sekanina, On an ordering of the set of vertices of a graph, Casopis Pést. Mat. 88 (1963) 265-282.

[39] P. Turán, Eine Extermalaufgabe aus der Graphentheorie, Mat. Fiz. Lapok 48 (1941), 436-452; On the theory of graphs, Colloq. Math. 3 (1954), 19-30.

[40] W. T. Tutte, On hamilton circuits, J. London Math. Soc. 21 (1946), 98-101.

[41] W. T. Tutte, The factorization of linear graphs, J. London Math. Soc. 22 (1947), 107-111.

[42] W. T. Tutte, The factors of graphs, Canad. J. Math. 4 (1952), 314-328.

[43] W. T. Tutte, A theorem on planar graphs, Trans. Amer. Math. Soc. 82 (1956), 99-116.

[44] M. E. Watkins and D. M. Mesner, Cycles and connectivity in graphs, Canad. J. Math. 19 (1967), 1319-1328.

[45] H. Whitney, A theorem on graphs, Ann. Math. 32 (1931), 378-390.

[46] M. Las Vergnas, Sur une propriété des arbres maximaux dans un graphe, C. R. Acad. Sci. Paris, A 272 (1971), 1297-1300.

[47] V. Chvátal, Edmonds polyhedra and a hierarchy of combinatorial problems, Discrete Math., to appear.

ON THE USE OF GRAPHS IN GROUP THEORY

Marshall D. Hestenes[1]

1. Introduction.

In recent years graph theory has become a very useful tool in the study of finite groups of small rank. This paper describes one method for obtaining graphs from permutation groups which are transitive on a set X. In the case when the groups have even order and are of rank 3 on X, the corresponding graphs are called rank 3 graphs and are contained in the class of strongly regular graphs. In order to provide those interested in strongly regular graphs with more examples and to give an indication of which strongly regular graphs are rank 3, we include a brief description of a few of the infinite families of rank 3 graphs. By far the most important contribution graph theory has made to group theory has been in the recent discovery of new finite simple groups. The central topic of this paper is the graph construction used to obtain

[1]Work supported in part by a grant from the National Science Foundation.

these groups.

2. Groups and their graphs.

Let G be a transitive permutation group acting on a finite set X, and denote the action of G on X by $x \to x^g$ for $x \in X$, $g \in G$. The group G also acts on $X \times X$ by $(x,y)^g = (x^g,y^g)$, and we call a G-orbit in $X \times X$ an orbital. Clearly, $I = \{(x,x) : x \in X\}$ is always an orbital of G. If Δ is an orbital, then for $x \in X$ set $\Delta(x) = \{y \in X : (x,y) \in \Delta\}$. Let G_x be the subgroup of G which fixes x, and observe that $\Delta(x)$ is an orbit of G_x on X as the following demonstrates. If $y \in \Delta(x)$ and $g \in G_x$, then $(x,y) \in \Delta$ implies $(x^g,y^g) = (x,y^g) \in \Delta$, and thus $y^g \in \Delta(x)$. Conversely, if y_1 and y are in the same G_x-orbit of X, then there exists a permutation $g \in G_x$ such that $y_1 = y^g$, so $(x,y) \in \Delta$ implies $(x^g,y^g) = (x,y_1) \in \Delta$ so $y_1 \in \Delta(x)$. Hence $\Delta(x)$ is a G_x-orbit of X. Now let Y be any G_x-orbit of X. If $x \in Y$, then $Y = \{x\}$. If $y \in Y$, $y \neq x$, then (x,y) is in some orbital $\Delta \neq I$ of $X \times X$, so $\Delta(x)$ is a G_x-orbit and $y \in \Delta(x)$, therefore $Y = \Delta(x)$. For fixed x, this establishes a one-to-one correspondence between G-orbitals and G_x-orbits! The rank of G is the number of orbitals of G, or equivalently the number of orbits of G_x for any x.

If Δ is an orbital, then $\Delta(x)^g = \Delta(x^g)$ for all $g \in G$. Moreover, the converse $\Delta^c = \{(y,x):(x,y)\in\Delta\}$ of Δ is again an orbital, so that $\Delta \to \Delta^c$ is a pairing of the set of orbitals, see Wielandt [14, p. 45].

Lemma. An orbital Δ is either symmetric ($\Delta = \Delta^c$) or asymmetric ($\Delta \cap \Delta^c = \emptyset$).

Proof. To see this, suppose Δ is not asymmetric and $\Delta \neq I$ (clearly I is symmetric). Then there must be two ordered pairs $(x_1,y_1),(y_1,x_1) \in \Delta$, with $x_1 \neq y_1$. Now let (x,y) be any other ordered pair in Δ. Since Δ is an orbital, there exist $g_1, g_2 \in G$ such that $(x_1,y_1) = (x,y)^{g_1}$ and $(y_1,x_1) = (x_1,y_1)^{g_2}$. Then

$$(x,y)^{g_1 g_2 g_1^{-1}} = (y_1,x_1)^{g_1^{-1}} = (y_1^{g_1^{-1}}, x_1^{g_1^{-1}}) = (y,x),$$ so Δ is symmetric.

The classical condition for the existence of a symmetric orbital is now stated. For the proof, see Wielandt [14, p. 45].

Theorem. A transitive permutation group G has a symmetric orbital other than I if and only if G has even order.

Since an orbital $\Delta \neq I$ is a relation on X, we can form a digraph (which when symmetric is a graph) $G_\Delta = (X,\Delta)$ having **vertex** set X and edge set (strictly speaking, directed edge set) Δ. Every orbital $\Delta \neq I$ is irreflexive, hence G_Δ has no loops and is either an ordinary graph (if symmetric) or an oriented graph (if asymmetric).

Using the 1-1 correspondence between orbitals of G and G_x-orbits, we now obtain an alternate description of the graph we have just constructed. Let x be a fixed element of X and let $\Delta(x)$ be an orbit of G_x on X corresponding to an orbital Δ. It is then obvious that $\Delta(x)$ consists of precisely the vertices of G_Δ that are adjacent to x. Since G is transitive, for all $y \in X$, there is a $g \in G$ such that $y = x^g$. But earlier we observed that $\Delta(x^g) = \Delta(x)^g$; hence the vertices adjacent to G_Δ to $y = x^g$ are the images under g of the vertices adjacent to x. This means that G is a subgroup of the automorphism group of G_Δ which is transitive on the edges of G_Δ as well as the vertices. It follows that G_Δ could have been constructed by joining a vertex x to a vertex y in the orbit $\Delta(x)$

of G_x; all other edges in G_Δ are images of this edge under elements of G.

<u>An easy example</u>. We illustrate the rank of a group using the cyclic group of order 4. Let $X = \{x_1,x_2,x_3,x_4\}$ and let $G = \langle(x_1\ x_2\ x_3\ x_4)\rangle$. Let Δ be the orbital containing (x_1,x_2). Then $\Delta = \{(x_1,x_2),(x_2,x_3),(x_3,x_4),(x_4,x_1)\}$ and G_Δ is an oriented graph consisting of the directed 4-circuit in Figure 1a.

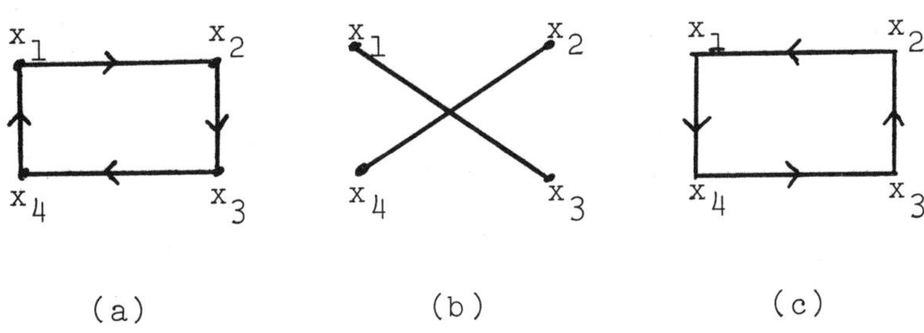

<u>Figure 1</u>

Let Γ be the orbital containing (x_1,x_3). Then $\Gamma = \{(x_1,x_3),(x_2,x_4),(x_3,x_1),(x_4,x_2)\}$ and so G_Γ is the (undirected) graph of Figure 1b. Let Φ be the orbital containing (x_1,x_4). Then $\Phi = \{(x_1,x_4),(x_2,x_1),(x_3,x_2),(x_4,x_3)\}$ and G_Φ is the oriented graph of Figure 1c which is the converse of G_Δ. The four orbitals Δ,Γ,Φ and I exhaust $X \times X$; hence this group G has rank 4.

Assume now that G is a rank 3 group with orbitals I,Δ, and Γ. We also suppose that G has even order, so by Wielandt's theorem Δ, and hence Γ are both symmetric; hence G_Δ and G_Γ are complementary graphs. We call a graph G_Δ obtained this way a rank 3 graph.

Remark. If a group G has rank 3, then let G* be the automorphism group of G_Δ. Since G $\subseteq$ G*, clearly G* is also transitive. Obviously, G* must preserve neighborhoods in G_Δ, so G* is also a rank 3 group. Hence a graph G is a rank 3 graph if and only if its automorphism group has rank 3.

By definition, a graph G is strongly regular if the neighborhoods N(x) consisting of all vertices of adjacent to x satisfy:

(1) $|N(x)|$ is independent of x (that is, the graph G is regular).

(2) $|N(x) \cap N(y)|$ for $x \neq y$ depends only on whether or not x and y are adjacent in G.

Theorem. Every rank 3 graph is strongly regular.

Proof. Notice that for a rank 3 graph, G_Δ, $N(x) = \Delta(x)$. Let x_1, x_2 be two elements of X.

Since G is transitive, there exists $g \in G$ such that $x_2 = x_1^g$; hence $\Delta(x_2) = \Delta(x_1^g) = \Delta(x_1)^g$. Since a permutation is one-to-one, $|\Delta(x_2)| = |\Delta(x_1)|$, and (1) is verified. To verify (2), suppose that x_1 is adjacent to y_1 and x_2 is adjacent to y_2 in G_Δ. Then we have $(x_1,y_1) \in \Delta$ and $(x_2,y_2) \in \Delta$. Since Δ is an orbital, there exists $g \in G$ such that $(x_2,y_2) = (x_1,y_1)^g$; hence

$$|\Delta(x_1) \cap \Delta(y_1)| = |\Delta(x_1)^g \cap \Delta(y_1)^g| = |\Delta(x_1^g) \cap \Delta(y_1^g)|$$
$$= |\Delta(x_2) \cap \Delta(y_2)|.$$

Therefore, $|\Delta(x) \cap \Delta(y)|$ is the same for all x adjacent to y.

Now suppose x_1 is not adjacent to y_1 and x_2 is not adjacent to y_2, where $x_1 \neq y_1$ and $x_2 \neq y_2$. Since G has rank 3, we have $(x_1,y_1),(x_2,y_2) \in \Gamma$, the other orbital different from I. Then Γ an orbital again implies there exists $g \in G$ such that $(x_2,y_2) = (x_1,y_1)^g$. The same computation as above gives $|\Delta(x_1) \cap \Delta(y_1)| = |\Delta(x_2) \cap \Delta(y_2)|$, so (2) is verified.

3. Parameters for rank 3 graphs.

If G is a rank 3 group with orbitals I, Δ and Γ, and $x \in X$, then $I(x) = \{x\}$, $\Delta(x)$ and $\Gamma(x)$ are the orbits of G_x in X. The lengths of these orbits: $1 = |I(x)|$, $k = |\Delta(x)|$, and $\ell = |\Gamma(x)|$ are called the <u>subdegrees</u> of G. Notice that $n = 1+k+\ell = |X|$ is the degree of the group G. Following [5], the intersection numbers, λ,μ of a rank 3 group G are defined by

$$|\Delta(x) \cap \Delta(y)| = \begin{cases} \lambda \text{ if } y \in \Delta(x) \\ \mu \text{ if } y \in \Gamma(x) \end{cases}$$

Of course the numbers n,k,ℓ,λ,μ are parameters of the rank 3 graph $\mathcal{G} = \mathcal{G}_\Delta$. The theory of strongly regular graphs usually denotes these parameters as follows: $n_1 = k$, $n_2 = \ell$, $p_{11}^1 = \lambda$ and $p_{11}^2 = \mu$.

From [4], we have the following development.

<u>Theorem</u>. Let A be the adjacency matrix of $\mathcal{G}$, a rank 3 graph. Then

$$(1) \qquad A^2 - (\lambda-\mu)A - (k-\mu)I = \mu J,$$

where I is the $n \times n$ identity matrix and J is the $n \times n$ matrix of all ones.

<u>Proof</u>. Recall that the (i,j)-entry of A^2 is the number of walks of length 2 from the vertex x_i to vertex x_j. Hence

$$(A^2)_{ij} = |\Delta(x_i) \cap \Delta(x_j)| = \begin{cases} k \text{ if } x_i = x_j \\ \lambda \text{ if } x_i \text{ is adjacent to } x_j \\ \mu \text{ if } x_i \text{ is not adjacent to } x_j. \end{cases}$$

It follows that

$$(A^2)_{ij} - (\lambda-\mu)(A)_{ij} - (k-\mu)(I)_{ij} = \begin{cases} k-0-(k-\mu)=\mu \text{ if } x_i = x_j \\ \lambda-(\lambda-\mu)-0=\mu \text{ if } x_i \text{ is adjacent to } x_j \\ \mu-0-0=\mu \text{ if } x_i \text{ is not adjacent to } x_j. \end{cases}$$

Therefore $A^2 - (\lambda-\mu)A - (k-\mu)I = \mu J$.

Since G is regular, the row sums of A are all $|\Delta(x)| = k$, so that the row sums of $(A-kI)$ are all 0 and $(A-kI)J = 0$. Hence A satisfies the polynomial equation

$$(x-k)(x^2-(\lambda-\mu)x-(k-\mu)) = 0.$$

Thus the spectrum of A consists of just three eigenvalues $\{k,r,s\}$ where

(2) $$r,s = \frac{\lambda-\mu \pm \sqrt{d}}{2} \text{ where } d = (\lambda-\mu)^2 + 4(k-\mu).$$

If G is connected, then by the Perron-Frobenius Theorem, k is a simple eigenvalue of A. If G is strongly regular and not connected, then all its components are complete and of the same size so $G = \frac{n}{k+1}K_{k+1}$, a collection of n/(k+1) disjoint copies of the complete graph on (k+1) vertices which is of course regular of degree k. Hence the intersection numbers are $\lambda = k-1$ and $\mu = 0$, and by equation (2), r = k and s = -1.

If we look only at connected graphs, the multiplicities f of r and g of s are determined by

$$n = 1 + f + g$$
$$0 = k + rf + sg \text{ (trace } A = 0)$$

to be

$$f = \frac{(n-1)s+k}{s-r} \qquad g = \frac{(n-1)r+k}{r-s} .$$

Since f and g must be integers, very strong necessary conditions are thus found for the parameters n,k,ℓ,λ,μ to belong to a rank 3 graph. Call the 9-tuple

$$(n,k,\ell,\lambda,\mu,r,s,f,g)$$

<u>the parameters</u> of a rank 3 graph.

The first five parameters of a rank 3 graph G are not independent, for clearly $n = 1+k+\ell$. In addition, k,ℓ,λ,μ are related by the equation $k(k-\lambda-1) = \ell\mu$. In order to see this, let x be a vertex of G and count the number of edges between $\Delta(x)$ and $\Gamma(x)$ in two ways. Every vertex in G has valency k. If y is a vertex in $\Delta(x)$, then there is an edge from y to x. The number of edges from y that go to vertices in $\Delta(x)$ is precisely the number of triangles containing the edge (x,y), namely λ. Therefore, for each of the k vertices in $\Delta(x)$, there are $k-\lambda-1$ edges going to vertices in $\Gamma(x)$, a total of $k(k-\lambda-1)$ edges in all. On the other hand, for each of the ℓ vertices z in $\Gamma(x)$, the number of edges to $\Delta(x)$ is just the number of $K_{1,2}$'s containing the non-edge {x,z}; hence there are $\ell\mu$ in all. Consequently, $k(k-\lambda-1) = \ell\mu$.

For further relations among the parameters, see [4].

4. Examples of families of rank 3 graphs.

Each of the known infinite families of rank 3 groups give rise to a corresponding family of rank 3 graphs by the above construction. Five of these families have been chosen as examples because they give the general flavor, and some of the graphs are well known and some are probably not. For each example a description of the group and the corresponding graph is given together with the parameters

$(n,k,\ell,\lambda,\mu,r,s,f,g)$.

Example (1). Let X be the set of all two-element subsets of $\{1,2,\ldots,m\}$, $m \geq 4$. The symmetric group $G = S_m$ on m letters acts as a transitive permutation group on X. If $x = (a,b) \in X$, then G_x has precisely two orbits on $X - \{(a,b)\}$, namely $\Delta(x) = \{(a,c) : c \neq a,b\} \cup \{(b,c) : c \neq a,b\}$ and $\Gamma(x) = \{(c,d) : c \neq a,b \text{ and } d \neq a,b\}$. Hence the graph $\mathcal{G} = \mathcal{G}_\Delta$ has X as its vertex set, with two distinct elements of X being adjacent if they intersect. This is a familiar graph to graph theorists. It is the line graph of K_m and is usually called a triangular graph. When m = 5, the complement $\mathcal{G}_\Gamma$ of $\mathcal{G}_\Delta$ is the Petersen graph. For m > 4, S_m is the automorphism group of $\mathcal{G}$ and is primitive, see Higman [6]. The parameters of $\mathcal{G}$ follow:

$$n = \binom{m}{2} \quad k = 2(m-2) \quad \lambda = m-2 \quad r = m-4 \quad f = m-1$$
$$\ell = \binom{m-2}{2} \quad \mu = 4 \quad s = -2 \quad g = \frac{m(m-3)}{2}$$

This graph $\mathcal{G}$ which is the line graph of K_m is the unique graph with these parameters for $m \neq 8$. For m = 8, there are exactly three exceptional graphs as shown by Chang and Hoffman [1,2,10], all of which have intransitive automorphism groups.

Example (2). This time let X be the set of all ordered pairs of m elements, $m \geq 2$, and let G be the wreath product $S_m[S_2]$ of the symmetric groups of degrees m and 2. Then G acts as a rank 3 permutation group on X. If $x = (a,b) \in X$, then

$\Delta(x) = \{(c,d) \in X - \{(a,b)\} : c = a \text{ or } d = b\}$.

Hence the graph $G = G_\Delta$ has vertex set X, and if these vertices are put in a matrix, each vertex is adjacent to the remaining vertices in the same row or column. This is the Latin Square graph $LS_2(m)$ and is the line graph of the complete bipartite graph $K_{m,m}$ also known as the Cartesian product graph $\overline{K}_m \times \overline{K}_m$.

$$n = m^2 \quad k = 2(m-1) \quad \lambda = m-2 \quad r = m-2 \quad f = 2(m-1)$$
$$\ell = (m-1)^2 \quad \mu = 2 \quad s = -2 \quad g = (m-1)^2$$

The graph G is the unique graph with these parameters for $m \neq 4$. For $m = 4$, there is one other graph as shown by Shrikhande [12], but it has rank greater than 3.

Example (3). Let $q = 4t + 1$ where q is a prime power and let $X = F_q$, the Galois field on q elements. The group $\{F_q,+\}$ acts regularly on X by $x^a = x + a$ for $x \in X$, $a \in \{F_q,+\}$. If $F_q^* = F_q - \{0\}$

and $\{F_q^*,\cdot\} = \langle\zeta\rangle$, then $\langle\zeta^2\rangle$ acts on X by $x^a = xa$ for $x \in X$, $a \in \langle\zeta^2\rangle$. Then $\langle\zeta^2\rangle$ has three orbits on X : {0}, the squares of F_q and the non-squares of F_q. The group G generated by the actions of $\{F_q,+\}$ and $\langle\zeta^2\rangle$ as described above has rank 3. The graph G_Δ has vertex set X, two vertices being adjacent if they differ by a non-zero square, see Higman [8]. This graph is self-complementary. The parameters follow

$$n = q \quad k = 2t \quad \lambda = t-1 \quad r = (-1+\sqrt{q})/2 \quad f = 2t$$
$$\ell = 2t \quad \mu = t \quad s = (-1-\sqrt{q})/2 \quad g = 2t.$$

<u>Example (4)</u>. Let $m \geq 4$ and let X be the set of lines of (m-1)-dimensional projective space $P_{m-1}(q)$ over F_q. Then the projective special linear group $PSL_m(q)$ acts as a rank 3 permutation group on X. Here G is the graph having vertex set X, with two lines being adjacent if they intersect. The full automorphism group of G is $P\Gamma L_m(q)$ which contains the projective general linear group if $m > 4$, while if $m = 4$, it is the group generated by $P\Gamma L_m(q)$ and the dualities of $P_3(q)$. Let $Q_m = (q^{m-1})/(q-1)$; then the parameters are

$$n = Q_m Q_{m-1}/Q_2 \quad k = q\ Q_2 Q_{m-2} \quad \lambda = Q_{m-1} - 2 + q^2$$
$$\ell = q^4 Q_{m-2} Q_{m-3}/Q_2 \quad \mu = Q_2^2$$
$$r = q\ Q_{m-2} - Q_2 \quad f = q\ Q_{m-1}$$
$$s = -Q_2 \quad g = \frac{q^2 Q_m Q_{m-3}}{q+1}\ .$$

For all even $m \geq 6$ and all odd $m \geq 21$, any rank 3 permutation group with degree n and subdegrees k and ℓ as given above is isomorphic to a subgroup of $P\Gamma L_m(q)$; see Higman [7].

Example (5). Let $N = 2m \geq 4$ and let X be the points of the symplectic $P_{N-1}(q)$. The symplectic group $S_N(q)$ acts as a rank 3 permutation group on X. The graph G then has X as vertex set, with two points of $P_{N-1}(q)$ being adjacent if they are orthogonal. The parameters follow:

$$n=Q_N \quad k=Q_{N-1}-1 \quad \lambda=Q_{N-2}-2 \quad r=q^{m-1}-1 \quad f=(q^m+q)Q_m/2$$
$$\ell=q^{n-1} \quad \mu=Q_{N-2} \quad s=-(q^{m-1}+1) \quad g=q(q^m+1)Q_{m-1}/2$$

We have just seen five infinite families of rank 3 graphs. There are over 20 such families known as well as numerous isolated graphs. Most of the remaining families come from permutation representations of the classical groups of their subgroups, just as in Examples (4) and (5).

5. Rank 3 extensions.

A rank 3 extension of an abstract group H is a rank 3 group G with $G_x \cong H$. By far the most important result obtained by the combined use of graph theory and group theory is the discovery of new finite simple groups as rank 3 extensions of known groups. In order to understand how these discoveries took place, assume first that G is a rank 3 permutation group on X with orbitals I, Δ, and Γ, and $\mathcal{G}_\Delta$ is the graph of orbital Δ. For each $x \in X$, G_x acts transitively on $\Delta(x)$ and $\Gamma(x)$. Hence G_x is a vertex transitive automorphism group on the subgraphs of $\mathcal{G}_\Delta$ induced by $\Delta(x)$ and $\Gamma(x)$. Furthermore, G_x is also an (intransitive) automorphism group on the induced subgraph of $\mathcal{G}_\Delta$ on $\Delta(x) \cup \Gamma(x)$. Of course, this implies that if $g \in G_x$ and $a \in \Delta(x)$ is adjacent to $b \in \Gamma(x)$, then $a^g \in \Delta(x)$ is adjacent to $b^g \in \Gamma(x)$. The problem of rank 3 extensions of known groups amounts to recovering G from G_x and a knowledge of its action on the above subgraphs.

Suppose then that a known group H is a candidate for the stabilizer of a point in a proposed rank 3 group G. First we must somehow find two graphs $\mathcal{G}_1$ and $\mathcal{G}_2$ on each of which H has a representation as a vertex transitive group of automorphisms. Denote the vertex sets of G_1 and G_2 by X_1 and X_2, respectively, and let P be one additional vertex which we assume is fixed by H. We wish to construct a rank 3 graph $G = G_\Delta$ with vertex set $X = X_1 \cup X_2 \cup \{P\}$ in such a way that $\Delta(P) = X_1$, $\Gamma(P) = X_2$ and G_1 and G_2 are the subgraphs of induced by $\Delta(P)$ and $\Gamma(P)$ respectively. Set $k = |X_1|$, $\ell = |X_2|$ and $n = 1+k+\ell$. Since H is transitive on X_1 and on X_2, the graphs G_1 and G_2 are point-symmetric and hence are regular; denote the valency of G_1 by λ and that of G_2 by $k-\mu$. Of course, n,k,ℓ,λ,μ must satisfy the conditions suggested in Section 3 for the existence of a parameter set. We assume $\mu \neq 0$ or k, so that the graph G will be connected.

To carry out our construction, we take the graphs G_1 and G_2 and join the additional vertex P to every vertex of G_1. The only edges that remain to be added are the $k(k-\lambda-1) = \ell\mu$ edges between vertices of X_1 and vertices of X_2. From our counting argument, we know that this must be done so that each of the k vertices of X_1 is adjacent to exactly $k-\lambda-1$ vertices of X_2, and each of the ℓ

vertices of X_2 is adjacent to exactly μ vertices of X_1. Moreover, if H is going to be in the automorphism group of G, it must be in the automorphism group of the subgraph of G induced by $X_1 \cup X_2$; i.e., the vertex $a \in X_1$ is adjacent to $b \in X_2$ if and only if $a^h \in X_1$ is adjacent to $b^h \in X_2$ for all $h \in H$. From the last comment, it is obvious that a thorough knowledge of the action of H on G_1 and G_2 is used in trying to add the edges between X_1 and X_2, and it may not be possible at all.

Even if it is possible, the graph G just constructed may not be point-symmetric and consequently cannot be rank 3. However, if G is point-symmetric, it will be rank 3. To see this, let K be the automorphism group of G. Certainly K is transitive on X since G is point-symmetric. Since H is in K and H fixes P, we have H is a subgroup of K_P, and since H has three orbits $\{P\}$, X_1 and X_2 on X, it follows that K_P has at most three orbits on X. Since G has diameter 2, no further collapsing of orbits can occur, so that the group K has rank exactly 3. If K has a transitive subgroup G such that $G_P = H$, then G is the desired rank 3 extension of H.

Let us look at an easy example of the above construction. Suppose H = S_3 as an abstract group. We find that there are two graphs G_1 with $X_1 = \{1,2,3\}$ and $G_2 = C_6$ with $X_2 = \{a,b,c,d,e,f\}$ on which H acts transitively. Here is a table of these actions.

H	Action of H on G_1	Action of H on G_2
h_1	1	1
h_2	(1 2 3)	(a e c)(b f d)
h_3	(1 3 2)	(a c e)(b d f)
h_4	(1 2)	(a b)(c f)(d e)
h_5	(1 3)	(a f)(b e)(c d)
h_6	(2 3)	(a d)(b c)(e f)

We now add a new vertex P and consider the set $X = \{P,1,2,3,a,b,c,d,e,f\}$. Join P to each vertex in G_1. We have $k=3$, $\ell=6$, $n=10$, $\lambda=0$ and $k-\mu=2$ so that $\mu=1$. Hence each vertex in G_1 must be joined to $k-\lambda-1 = 2$ vertices in G_2 in such a way that each vertex in G_2 is adjacent to exactly one vertex in G_1. First let us determine which vertices in G_2 to join to 1. The stabilizer in H of 1 is $\{h_1,h_6\}$ and this has orbits $\{a,d\}$, $\{b,c\}$, and $\{e,f\}$ on the vertices in G_2. Let x and y denote the points in one of these orbits. If 1 is adjacent to x, then $1^{h_6} = 1$ must be adjacent to $x^{h_6} = y$. Thus 1 must be adjacent to both points in one orbit and nonadjacent to the remaining points. If 1 is adjacent to either b and c or to e and f, then since b is adjacent to c and e is adjacent to f, vertex 1 belongs to a triangle while the vertex P does not; thus G would not be point-symmetric. Hence 1 must be adjacent to a and d. Applying h_4 we see that $1^{h_4} = 2$ must be adjacent to $a^{h_4} = b$ and $d^{h_4} = e$. Similarly, 3 must be adjacent to c and f, as shown in Figure 2.

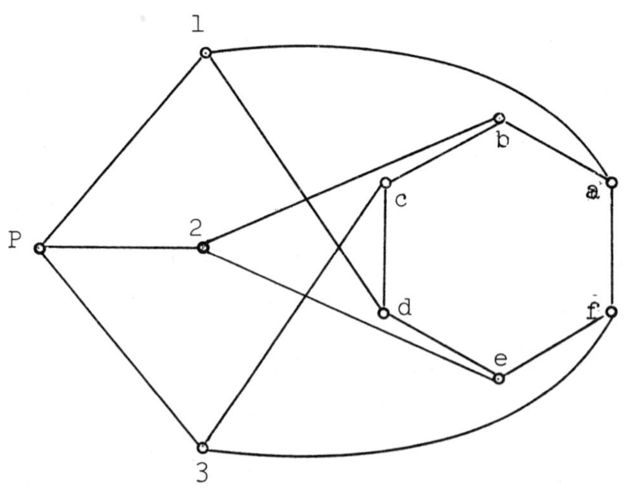

Figure 2

There are no triangles and each non-adjacent pair of vertices is connected by one path of length two, so that G is strongly regular. In fact, if we redraw G we see in Figure 3 that it is the very familiar Petersen graph.

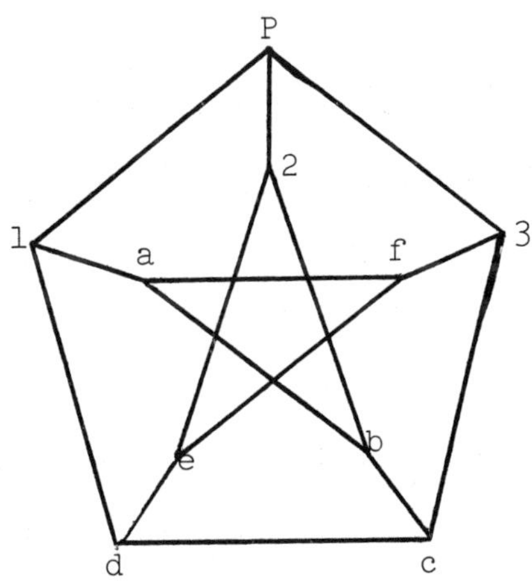

Figure 3

By Example (1) of Section 3, the automorphism group of G is isomorphic to S_5. The alternating group $G = A_5$ is a rank 3 subgroup of S_5 and $H = G_P$.

An another example, consider $H = A_5$. This time choose $G_1 = \overline{K}_5$ and G_2 = the Petersen graph of Figure 3. To display the simultaneous action of H on G_1 and G_2, represent G_1 as the graph with five isolated points {1,2,3,4,5}, and represent G_2 as the complement of the line-graph of K_5. As such, the vertices of G_2 can be thought of as the unordered pairs {12,13,14,15,23,24,25,34,35,45} where ij is adjacent to kℓ if and only if $\{i,j\} \cap \{k,\ell\} = \Phi$. Now construct G on the vertex set

$$\{P,1,2,3,4,5,12,13,14,15,23,24,25,34,35,45\}.$$

In addition to the edges already in G_1 and G_2, join P to each of {1,2,3,4,5} and join i to jk if and only if i = j or i = k. Clearly this method of joining {P}, G_1 and G_2 is consistent with the actions of H on G_1 and G_2 . We need only show that G is point-symmetric. The way G was defined, it is obvious that all the points of G_1 are symmetric with each other and the points of G_2 are symmetric with each other. We need only show that the graph has automorphisms carrying P to some point of G_1 and to some point of G_2. Consider the following drawing of G given by Figure 4.

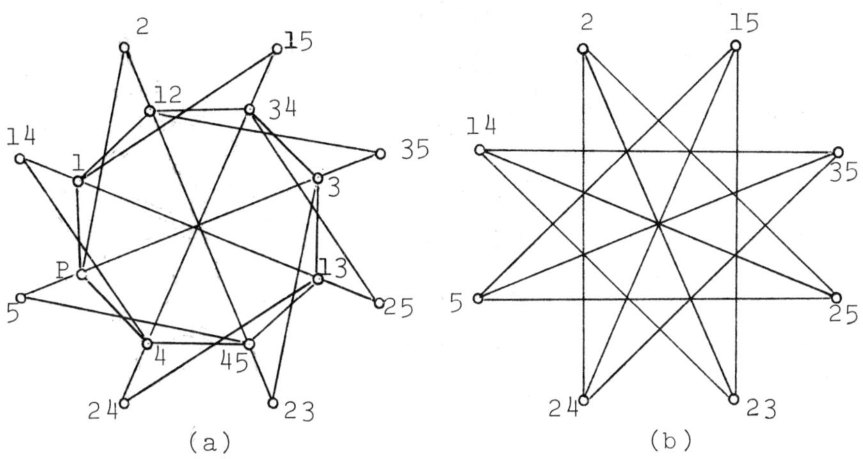

Figure 4

For clarity's sake, 12 edges of G have been left out of Figure 4a. Those 12 edges are between points of the outer ring as shown in Figure 4b, which is also known as the möbius ladder on 8 points. It is obvious that this drawing of G shows that there are automorphisms (corresponding to rotating the figure) that carry P to points of G_1 and to points of G_2, so G is a rank 3 graph. Notice that we still do not have a rank 3 extension of H until we find some subgroup of the automorphism group of G in which H is the point stabilizer. Set a = (2,3)(4,5)(12,13)(14,15)(24,35)(25,34),

b = (1,2)(3,4)(13,24)(14,23)(15,25)(35,45),

c = (1,3)(4,5)(12,23)(14,35)(15,34)(24,35), and

d = (P,45,3,12)(1,4,13,34)(2,5,23,35)(14,24,25,15).

Then H = <a,b,c> and d is the 90° rotation of G. Since d carries P to 3 ε G_1 and 45 ε G_2, the group G = <a,b,c,d> is transitive on G. A computer program quickly shows the order of G to be 960. Since G is transitive on the 16 points of X, the order of G_P is 960/16 = 60. But H is a subgroup of G_P of order 60; hence H = G_P and G is a rank 3 extension of H.

In this example, it is easy to see that G is a proper subgroup of the automorphism group K of G. Certainly S_5 acts on G_1 and on G_2, and these actions are consistent with the way we joined G_1 and G_2. Hence K_P contains a subgroup isomorphic to S_5 while certainly G_P does not.

6. The 4-vertex condition

Our attempt to find rank 3 graphs will be simplified if we can obtain more information about the structure of a rank 3 graph. We have already pointed out that a rank 3 graph G is point-symmetric and strongly regular. Since the edge set of G is an orbital for the automorphism group of , every rank 3 graph is also edge-symmetric.

Now a strongly regular graph is said to satisfy the 4-*vertex condition* if the number α of K_4's containing a given edge is independent of the edge and the number β of subgraphs of the type  x y is independent of $\{x,y\}$.

Theorem. Every rank 3 graph satisfies the 4-vertex condition.

Proof. Since the number of K_4's containing a given edge is invariant under automorphism, every rank 3 graph must satisfy the first criterion of the 4-vertex condition. Also, since $G_\Delta^c = G_\Gamma$ also has rank 3, the complement of every rank 3 graph is edge-symmetric. Hence all rank 3 graphs must satisfy the second criterion of the 4-vertex condition too.

Theorem. In a rank 3 graph G, the numbers α and β are related by $\ell\beta = k(\lambda(\lambda-1)/2-\alpha)$.

Proof. We count the number of induced subgraphs of the form J = (K_4 minus one edge) in two ways. Let P be a fixed vertex of G. If the non-edge of J contains P, then the other end of the non-edge must be in $\Gamma(P)$, the points of distance two from P, and hence there are ℓ choices for that vertex. By the definition of β, there are β subgraphs J containing a given non-edge. Consequently, there are $\ell\beta$ induced J's in G that contain P on a non-edge.

Now count the number of induced J's in G that contain P, where P is not on the non-edge. There are k choices for the other point not on the non-edge, and for each such $Q \in \Delta(P)$, we look for non-edges in $\Delta(P) \cap \Delta(Q)$. Since $|\Delta(P) \cap \Delta(Q)| = \lambda$, there are $\lambda(\lambda-1)/2$ unordered pairs of points in this set, and they form either edges or non-edges. The edges in the $\lambda(\lambda-1)/2$ possible pairs correspond to K_4's containing (P,Q), and there are α such K_4's; hence there are $\lambda(\lambda-1)/2 - \alpha$ non-edges, both of whose endpoints are adjacent to Q and P. Consequently, there are $k(\lambda(\lambda-1)/2-\alpha)$ induced subgraphs J containing P, but not on a non-edge.

Since each J has 2 points on the non-edge and 2 points not on the non-edge, the total number of J's in G is $n\beta\,\ell/2$ by the first count and $nk(\lambda(\lambda-1)/2-\alpha)/2$ by the second. Therefore, $\beta\ell = k(\lambda(\lambda-1)/2-\alpha)$.

Assume now that $G = G_\Delta$ is a rank 3 graph, P is a vertex of G, and the subgraph G_1 of G induced by $\Delta(P)$ is also a rank 3 graph. Immediately one observes that there are relations between the parameters n,k,ℓ,λ,μ etc. of G and the parameters $n_1,k_1,\ell_1,\lambda_1,\mu_1$ etc. of G_1. For instance, $k = |\Delta(P)|$, but G_1 is the subgraph induced by $\Delta(P)$ implies $n_1 = |\Delta(P)|$; hence $n_1 = k$. Similarly, if $P_1 \varepsilon \Delta(P)$, the number of triangles in G is λ, precisely the number k_1 of vertices of G_1 adjacent to P_1; hence $\lambda = k_1$. As a third relation, we have $\alpha = k_1\lambda_1/2$, for the number of K_4's with edge (P,P_1) is just the number of triangles in the rank 3 graph G_1 which contain the vertex P_1.

7. The search for new groups.

Recall that in attempting to construct a rank 3 extension G of a given group H, we needed to know two graphs G_1 and G_2 on which H was vertex transitive. The case that has been most fruitful to date occurs when the graph G_1 is itself rank 3. Suppose for a moment we are in this case and G_2 has not yet been found. By what we have just seen, the parameters of G_1 determine the parameters k,λ and α of G. A computer program has been implemented which determines the possible remaining parameters of G. The usual conditions on the parameters of strongly regular graphs suggested in Section 3 (see Hestenes and Higman [4, Section 4]) and the further conditions determined by the 4-vertex condition (see Hestenes and Higman [4, Section 5]) are imposed by the program. The computer output is somewhat surprising in that frequently there are no possible parameter sets for G and seldom are there more than one. If a parameter set exists, the graph G_2 must have ℓ vertices and valency $k - \mu$. If such a graph G_2 exists on which H is transitive, then we are ready to proceed with the construction described earlier.

The first new simple group obtained by the construction has degree 100 (i.e., the graph G has 100 vertices) and order 44,352,000, see Higman and Sims [9]. The group H used was M_{22}, the Mathieu group in its actions on 22 letters and on the 77 blocks of a Steiner system S(3,6,22). Other groups obtained this way are McLaughlin's group [11] with degree 275, order 898,128,000 and H = $U_4(3)$, Suzuki's group [13] with degree 1782, order 448,345,497,600 and H = $E_2(4)$, and B. Fischer's three groups, the smallest of which has degree 3510 and order about 64×10^{12}. In addition, two of the other new simple groups could have been obtained by our construction.

A frequently asked question is how does one know these new groups are simple? The answer is that when one starts with a group H which is simple, it is usually easy to determine if the extension G is simple. We need only the following theorems.

(1) A minimal normal subgroup of a group is characteristically simple (the direct product of isomorphic simple groups), see Gorenstein [3, p. 17].

(2) A nontrivial normal subgroup of a primitive group is transitive, see Wielandt [14, p. 13].

It is trivial to check whether the group G is primitive, for by a theorem in [4], the group G is primitive if both G and $\overline{G}$ are connected, i.e., $\mu \neq 0$ or k. Assume that G is a primitive group acting on a finite set X and G_P is simple for all $P \in X$. We will show that either G is simple or G contains a regular normal subgroup which is characteristically simple.

Let N be a nontrivial normal subgroup of G. Then for $P \in X$, $N \cap G_P = N_P$ is a normal subgroup of G_P; consequently $N_P = G_P$ or $N_P \cap G_P = 1$ since G_P is simple. If $N_P = G_P$, then since N is transitive on X by (2), $|N| = |X| \cdot |N_P| = |X| \cdot |G_P| = |G|$. Hence $N = G$, a contradiction. Therefore, $N_P \cap G_P = 1$ and N is a regular normal subgroup of G. Of course $|N| = |X|$. Any proper subgroup of N is not transitive on X, since the degree $|X|$ divides the order of a transitive group, so that by (2), N is a minimal normal subgroup of G. By (1), N is characteristically simple.

In the cases that have arisen, $n = |X|$ is not even the order of a characteristically simple group. Consequently, the groups must be simple.

8. Symmetric Block Designs (SBD's).

A knowledge of existing rank 3 graphs as well as the newly constructed ones may be useful in several areas of combinatorial theory. As one example, we include the observation made several times in the past of the connection between rank 3 graphs and SBD's.

Let $\mathcal{G}$ be a rank 3 graph. If $\lambda = \mu$, we can define an SBD on the point set X with blocks

$B_1,\ldots,B_n$ defined by $B_i = \Delta(x_i)$. These blocks have the right intersection property since

$$|\Delta(x_i) \cap \Delta(x_j)| = \begin{cases} \lambda \text{ if } x_i \text{ is adjacent to } x_j \\ \mu \text{ if } x_i \text{ is not adjacent to } x_j \end{cases}$$

and if $\lambda = \mu$, the number of elements of $B_i \cap B_j$ is a constant. Since $|B_i| = k$, we obtain an (n,k,λ) SBD. Similarly, suppose $\mu = \lambda+2$. We then can define blocks $B_1,\ldots,B_n$ by

$$B_i = \Delta(x_i) \cup \{x_i\}.$$

Notice that if x_i is adjacent to x_j,

$$|(\Delta(x_i)\cup\{x_i\})\cap(\Delta(x_j)\cup\{x_j\})| = |\{x_i\}\cup\{x_j\}\cup(\Delta(x_i)\cap\Delta(x_j))|$$
$$= \lambda + 2.$$

If x_i is not adjacent to x_j, then

$$|(\Delta(x_i)\cup\{x_i\})\cap(\Delta(x_j)\cup\{x_j\})| = |\Delta(x_i)\cap\Delta(x_j)| = \mu.$$

Therefore, if $\lambda + 2 = \mu$, we have an $(n,k+1,\mu)$ SBD.

Notice that the second example of Section 5 had parameters $n=16$, $k=5$, $\ell=10$, $\lambda=0$, $\mu=2$. Thus $\lambda + 2 = \mu$ and we have a $(16,6,2)$ SBD.

REFERENCES

[1] L. C. Chang, The uniqueness and nonuniqueness of the triangular association scheme, Sci. Record 3 (1959), 604-613.

[2] L. C. Chang, Association schemes and partially balanced block designs with parameters v=28, $n_1=12$, $n_2=15$ and $p_{11}^2=4$, Sci. Record 4 (1960), 12-18.

[3] D. Gorenstein, Finite Groups, Harper and Row, New York, 1968.

[4] M. D. Hestenes and D. G. Higman, Rank 3 groups and strongly regular graphs, Computers in Algebra and Number Theory, SIAM-AMS Proceedings 4 (1971), 141-159.

[5] D. G. Higman, Finite permutation groups of rank 3, Math. Z. 86 (1964), 145-156.

[6] D. G. Higman, Characterization of families of rank 3 permutation groups by the subdegrees I, Arch. Math. (Basel) 21 (1970), 151-156.

[7] D. G. Higman, Characterization of families of rank 3 permutation groups by the subdegrees II, Arch. Math. (Basel) 21 (1970), 353-361.

[8] D. G. Higman, Solvability of a class of rank 3 permutation groups, Nagoya Math. J. 41 (1970).

[9] D. G. Higman and C. C. Sims, A simple group of order 44,352,000, Math. Z. 105 (1968), 110-113.

[10] A. J. Hoffman, On the uniqueness of the triangular association scheme, Ann. Math. Statist. 31 (1960), 492-497.

[11] J. McLaughlin, A simple group of order 898,128,000, Theory of Finite Groups, (R. Brauer, ed.) Benjamin, New York (1969), 109-111.

[12] S. S. Shrikhande, The uniqueness of the L_2 association scheme, Ann. Math. Statist. 30 (1959), 781-798.

[13] M. Suzuki, A simple group of order 448,345,497,600, Theory of Finite Groups, (R. Brauer, ed.) Benjamin, New York (1969), 113-119.

[14] H. Wielandt, Finite Permutation Groups, Academic Press, New York, 1964.

THE ADJACENCY MATRIX AND THE GROUP OF A GRAPH*

Abbe Mowshowitz

1. Introduction.

Let G be a graph with automorphism group $\Gamma = \Gamma(G)$, and adjacency matrix $A = A(G)$. If σ is a permutation defined on the nodes of G, then $\sigma \in \Gamma$ if and only if P_σ, the permutation matrix corresponding to σ, commutes with A. This simple observation forms the basis of a useful connection between the automorphism group and the adjacency matrix of a graph. The problem of computing the automorphism group is seen to be equivalent to finding all permutation matrices which commute with the adjacency matrix. Conversely, to find all graphs with a given automorphism group (regarded as a group of permutation matrices), it suffices to determine all symmetric (0,1)-matrices with zero diagonal which commute with every permutation matrix in the given group and with no others.

*This research was supported in part by grant NRC A-7328 from the National Research Council of Canada.

The link between the group and the matrix of a graph was first exploited by Chao [2] to construct graphs whose automorphism groups contain a given transitive group as a subgroup. The construction is based on a result of Schur (see [12]), which guarantees that the set of all square matrices, with entries in a field F, which commute with each element of a given transitive group of permutation matrices forms a vector space over F. Moreover, a set of matrices which form a basis for the space can be constructed in terms of the orbits of the stabilizer of an element in the object set of the given group.

To construct the automorphism group $\Gamma = \Gamma(G)$ of a graph G using the adjacency matrix $A = A(G)$, one might first determine the class C of matrices which commute with A over a given field, and then select those elements of C which are permutation matrices. Using this approach, Mowshowitz [8] proved that if the eigenvalues of the adjacency matrix of a graph are pairwise distinct, then every non-trivial automorphism has order 2, so that the group is abelian, which in turn implies (by a result of Chao [1]) that for $p > 2$ no graph with p nodes satisfying this condition has a transitive group. Moreover, Chao [3] showed that under the same condition, the automorphism group of a directed graph is abelian. In both cases, the proof depends on the solution (see [5]) to the matrix equation

$$AX = XA \tag{1}$$

for all matrices X over the complex number field. When A has distinct eigenvalues, X is of the form

$$X = U \hat{X} U^{-1} \tag{2}$$

where $\hat{X}$ is a diagonal matrix whose diagonal entries are arbitrary parameters, and U is a non-singular matrix which transforms A into its Jordan canonical form. From this representation of X, one may readily infer the desired property of permutation matrices of the form $U \hat{X} U^{-1}$.

In what follows we will be concerned principally with the construction of the automorphism group of a graph or a digraph. Although most of the discussion will be seen to be applicable to nets, we will confine our attention to graphs and digraphs, inasmuch as the generalization to nets is immediate (see [10]). Since the adjacency matrix of a graph or digraph is a (0,1)-matrix, it may be regarded as a matrix over a finite field. This point of view facilitates the construction of automorphism groups, as will be demonstrated in the sequel.

Let $D = (V,X)$ be a digraph with node set $V = V(D) = \{v_1,v_2,\ldots,v_p\}$ and line set $X = X(D)$. Note that a graph is a symmetric digraph. The adjacency matrix $A = A(D) = (a_{ij})$ of D is defined by

$$a_{ij} = \begin{cases} 1 \text{ if } v_i \text{ is adjacent to } v_j \\ 0 \text{ otherwise} \end{cases}$$

The automorphism group $\Gamma = \Gamma(D)$ of a digraph D is the set of all one-one mappings of V onto itself which preserve adjacency. For graph theoretic terms not defined here, see [6],[7]. Throughout the following, we shall regard all groups as groups of permutation matrices.

2. Digraphs with non-derogatory adjacency matrix.

Let D be a p-node digraph with adjacency matrix $A = A(D)$. As we observed earlier, an element $P \varepsilon S_p$ (the symmetric group of degree p) is in $\Gamma(D)$ if and only if

$$PA = AP \tag{3}$$

Thus, taking A as a matrix over a field F, $\Gamma(D)$ is contained in the centralizer of A over F.

<u>Theorem 1</u>. Let D be a digraph with adjacency matrix A = A(D). If A is non-derogatory (i.e., its minimal and characteristic polynomials are identical) over a field F, then Γ(D) is abelian.

<u>Proof</u>. Since A is non-derogatory, the centralizer of A is just the ring of polynomials in A over F (see, for example, [11]). Thus, every P ε Γ(D) is a polynomial in A, from which the result follows.

<u>Corollary 1a</u>. (Chao [3]) If the adjacency matrix of a digraph D has all distinct eigenvalues, Γ(D) is abelian.

<u>Proof</u>. Since A(D) has distinct eigenvalues, it is non-derogatory over the field of complex numbers.

<u>Corollary 1b</u>. (Mowshowitz [8]) If the adjacency matrix A = A(G) of a graph G has all distinct eigenvalues, then Γ(G) is abelian and every non-trivial automorphism has order 2.

<u>Proof</u>. First, we observe that A is a symmetric matrix, so that all of its eigenvalues are real. By hypothesis, A is non-derogatory over the reals. Hence, Γ(G) is abelian. Moreover, since each polynomial in A over the reals is a symmetric matrix, it is clear that all non-trivial automorphisms have

order 2.

3. Group structure and the factorization of the characteristic polynomial.

An important special case of Theorem 1 arises when the characteristic polynomial is irreducible over the integers. The following result is a generalization of some observations of Collatz and Sinogowitz [4].

Theorem 2. Let D be a digraph with $A = A(D)$, $\Gamma = \Gamma(D)$, and characteristic polynomial $\phi_A(x)$ of degree p; and let k be the number of orbits of Γ. Then there exists a polynomial of degree k dividing $\phi_A(x)$.

Proof. Without loss of generality, suppose the rows of A are arranged into blocks corresponding to the orbits of Γ. Now, let $\zeta = [z_1,\ldots,z_1\ldots z_k,\ldots,z_k]^T$ be a vector with h_i components equal to z_i where h_i is the size of the i-th orbit of Γ. Consider the result of multiplying ζ by A.

$$(4)\quad A\zeta = [\sum_{j=1}^{k} z_j t_{1j},\ldots,\sum_{j=1}^{k} z_j t_{1j},\ldots,\sum_{j=1}^{k} z_j t_{kj},\ldots,\sum_{j=1}^{k} z_j t_{kj}]^T$$

where t_{ij} is the number of lines incident from a

node in the i-th orbit to nodes in the j-th orbit of Γ. (Clearly, the number of such lines is the same for all nodes in the same orbit.)

Now, let $T = (t_{ij})$ for $1 \leq i, j \leq k$, $\bar{\zeta} = [z_1, \ldots, z_k]^T$, and consider the equation

$$A\zeta = x\zeta. \tag{5}$$

From (4) and (5) we obtain

$$T\bar{\zeta} = x\bar{\zeta}. \tag{6}$$

Hence, $\bar{\zeta}$ is an eigenvector of T, and consequently ζ is an eigenvector of A, so that det(T-xI), the characteristic polynomial of T, divides $\phi_A(x)$, which concludes the proof.

Corollary 2a. If $\phi_A(x)$ is irreducible over the integers, then Γ is trivial.

Figure 1 exhibits a smallest digraph and graph with irreducible characteristic polynomials.

$x^3 - x - 1$

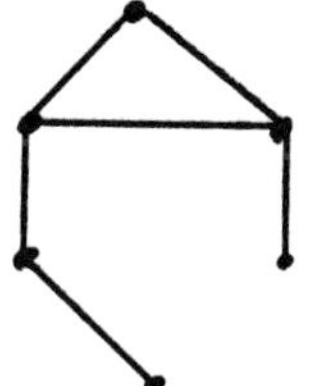

$x^6-6x^4-2x^3+7x^2+2x-1$

Figure 1. Digraphs with irreducible polynomial

To illustrate the proof of Theorem 2, we will factor the characteristic polynomial of the adjacency matrix of the graph G shown in Figure 2.

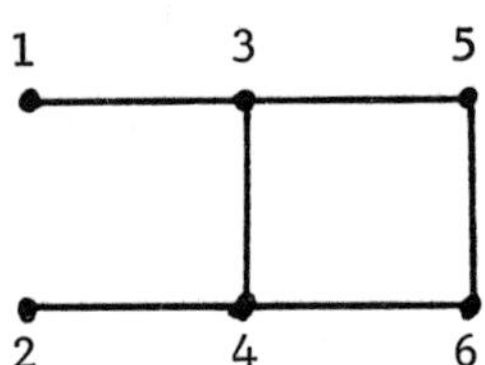

Figure 2. A graph G whose characteristic polynomial factors into two irreducible polynomials

The characteristic polynomial of $A = A(G)$ is $\phi_A(x) = x^6 - 6x^4 + 5x^2 - 1$. Since $\Gamma(G)$ is non-trivial, Theorem 2 guarantees that $\phi_A(x)$ has a non-trivial factorization. The orbits of $\Gamma(G)$ are clearly $\{1,2\}$, $\{3,4\}$, and $\{5,6\}$, so if $\zeta = [z_1,z_1,z_2,z_2,z_3,z_3]^T$, then $A\zeta$ becomes

$$\begin{pmatrix} 001000 \\ 000100 \\ 100110 \\ 011001 \\ 001001 \\ 000110 \end{pmatrix} \cdot \begin{pmatrix} z_1 \\ z_1 \\ z_2 \\ z_2 \\ z_3 \\ z_3 \end{pmatrix} = \begin{pmatrix} z_2 \\ z_2 \\ z_1+z_2+z_3 \\ z_1+z_2+z_3 \\ z_2+z_3 \\ z_2+z_3 \end{pmatrix}$$

The matrix T in this example is given by

$$T = \begin{pmatrix} 010 \\ 111 \\ 011 \end{pmatrix}$$

and $\bar{\zeta} = [z_1,z_2,z_3]^T$. Now, the equation $A\zeta = x\zeta$

reduces to $T\bar{\zeta} = x\bar{\zeta}$ and

$$\phi_T(x) = \det(xI-T) = \det \begin{pmatrix} x & -1 & 0 \\ 1 & x-1 & -1 \\ 0 & -1 & x-1 \end{pmatrix}$$

$$= (x^3 - 2x^2 - x + 1).$$

Hence, $\phi_A(x) = (x^3 - 2x^2 - x + 1)(x^3 + 2x^2 - x - 1)$.

The converse of Theorem 2 is not true, as evidenced by the fact that the characteristic polynomials of trees with an odd number of nodes, and of regular graphs have linear factors.

Corollary 2b. Let $\phi_A(x) = \alpha(x)\beta(x)$, where $\alpha(x)$ and $\beta(x)$ are irreducible of degrees m and n, respectively. Then the number of orbits of Γ is m,n, or m+n.

Thus, for example, if a graph G has at least three nodes and $\phi_{A(G)}(x) = (x-a)\alpha(x)$ where $\alpha(x)$ is irreducible, then $|\Gamma(G)| \leq 2$. This follows from the fact that $\Gamma(G)$ is abelian since A(G) is non-derogatory and thus $\Gamma(G)$ cannot be transitive, for $\Gamma(G)$ cannot be both abelian and transitive (Chao [1]). Of the eight identity graphs on six nodes, only one (shown in Figure 3) has a reducible characteristic polynomial.

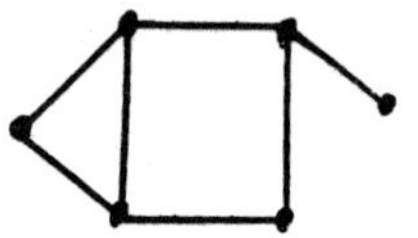

$$x(x^5-8x^3-6x^2+8x+6)$$

Figure 3. An identity graph with reducible polynomial

4. Construction of the group of a graph with non-derogatory matrix*

Theorem 1 implies that every non-trivial automorphism of a graph with non-derogatory adjacency matrix (with respect to a given field) has order 2. This follows from the fact that any polynomial in a symmetric matrix is again a symmetric matrix. For purposes of constructing the group of a graph, it is useful to regard the adjacency matrix as one with entries in GF(2). So, let G be a p-node graph with group $\Gamma = \Gamma(G)$ and non-derogatory adjacency matrix $A = A(G)$ over GF(2).

Since any matrix Q satisfying $AQ = QA$ is a polynomial in A, we can write

*A list of all connected graphs with up to five nodes together with their characteristic and minimal polynomials (with respect to GF(2)) is given in an appendix.

(7) $$Q = \sum_{i=0}^{p-1} a_i A^i \quad , \quad a_i \in GF(2)$$

Now, if Q is a permutation matrix (and thus an element of Γ) we have

(8) $$I = Q^2 = [\sum_{i=0}^{p-1} a_i A^i]^2 = \sum_{i=0}^{p-1} a_i (A^2)^i$$

Thus, in order to find the elements of Γ, it suffices to examine all polynomials f(x) such that $f(A^2) = I$; for, if f(A) is a permutation matrix, $f(A) \in \Gamma$.

Let $\mu_{A^2}(x)$ denote the minimal polynomial of A^2, and suppose $\deg \mu_{A^2}(x) = m$. Then every polynomial f(x) satisfying $f(A^2) = I$ is of the form

(9) $$f(x) = g(x)\, \mu_{A^2}(x) + 1$$

for some polynomial g(x). Hence, every matrix Q such that $Q^2 = I$ can be expressed in the form

(10) $$Q = I + \mu_{A^2}(A) \sum_{i=0}^{p-m-1} b_i A^i$$

where $b_i \in GF(2)$.

<u>Lemma 3a</u>. Let A = A(G) be the adjacency matrix of a p-node graph G. Suppose A is non-derogatory over GF(2). Then

$$[\mu_{A^2}(x)]^2 = \mu_{A^2}(x^2) = \begin{cases} \phi_A(x) \text{ if p is even} \\ x\ \phi_A(x) \text{ if p is odd.} \end{cases}$$

Proof. First, let us regard the coefficients of $\phi_A(x)$, the characteristic polynomial of A, as integers.

$$\phi_A(x) = \sum_{i=0}^{p} (-1)^i a_i\ x^{p-i}$$

By Theorem 2 of [9], all the odd subscripted coefficients a_i are even. Hence if p is even

$$\phi_A(x) = \sum_{i \text{ even}} a_i\ x^{p-i} = \sum_{i \text{ even}} a_i\ (x^2)^{\frac{p-i}{2}} ;$$

if p is odd

$$\phi_A(x) = x \sum_{i \text{ even}} a_i (x^2)^{\frac{p-i-1}{2}}$$

Once again, regarding $\phi_A(x)$ as a polynomial over GF(2), we see that

$$\phi_A(c) = \begin{cases} [\sum_{i \text{ even}} a_i\ x^{\frac{p-i}{2}}]^2 \text{ if p is even} \\ x[\sum_{i \text{ even}} a_i\ x^{\frac{p-i-1}{2}}]^2 \text{ if p is odd} \end{cases}$$

Now, suppose p is even and let $g(x) = \sum_{i \text{ even}} a_i\ x^{\frac{p-i}{2}}$

Clearly, $g(A^2) = 0$. If h(x) is a polynomial of degree $< \frac{p}{2}$ and $h(A^2) = 0$, then $h^2(x)$ is such that

$h^2(A) = 0$ and deg $h^2(x) < p$, contradicting the minimality of p. The argument is exactly analogous for p odd.

From Lemma 3a and the foregoing discussion, we obtain the following.

Theorem 3. Let G be a p-node graph with adjacency matrix $A = A(G)$. If A is non-derogatory over GF(2), then every automorphism $P \in \Gamma(G)$ can be expressed in the form

$$P = \mu_{A^2}(A) \sum_{i=0}^{p-m-1} b_i A^i + I \tag{11}$$

for some choice of $b_i \in GF(2)$, where $m = \deg \mu_{A^2}(x) = \{\frac{p}{2}\}$. Thus, $\Gamma(G)$ can be constructed in at most $2^{[\frac{p}{2}]}$ steps.

As an application of Theorem 3, let us construct the group of the graph shown in Figure 4.

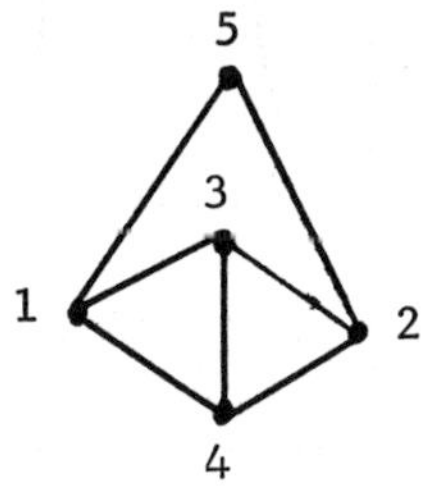

Figure 4. A graph G with non-derogatory adjacency matrix over GF(2)

The characteristic polynomial of $A = A(G)$ is

$$\phi_A(x) = x^5-7x^3-4x^2+2x = (x^3-x^2-6x+2)\ x(x+1)$$

With respect to GF(2), $\phi_A(x) = x^3(x+1)^2$ and it is easily verified that $\phi_A(x)$ is non-derogatory over GF(2). The minimal polynomial of A^2 over GF(2) is $\mu_{A^2}(x) = x^2(x+1)$ so that $m = \deg \mu_{A^2}(x) = 3$ and $n-m-1 = 1$. So, if P is an automorphism of G,P can be expressed in the form

$$\begin{aligned} P &= \mu_{A^2}(A)\ [b_0 I + b_1 A] + I \\ &= (A^3 + A^2)\ (b_0 I + b_1 A) + I \end{aligned} \tag{12}$$

Substituting for b_0 and b_1 in (12), we obtain the matrices I, $A^3 + A^2 + I$, $A^4 + A^3 + I$, and $A^4 + A^2+I$. Computing successive powers of A.

$$A = \begin{pmatrix} 00111 \\ 00111 \\ 11010 \\ 11100 \\ 11000 \end{pmatrix};\ A^2 = \begin{pmatrix} 11110 \\ 11110 \\ 11100 \\ 11010 \\ 00000 \end{pmatrix};\ A^3 = \begin{pmatrix} 00110 \\ 00110 \\ 11010 \\ 11100 \\ 00000 \end{pmatrix};\ A^4 = \begin{pmatrix} 00110 \\ 00110 \\ 11100 \\ 11010 \\ 00000 \end{pmatrix}$$

one finds $A^3 + A^2 + I = P_{(12)(34)}$, $A^4 + A^3 + I=P_{(34)}$ and $A^4 + A^2 + I = P_{(12)}$, so that in this example all $2^{[5/2]} = 4$ different matrices of the form

$\mu_{A^2}(A) \sum_{i=0}^{1} b_i A^i + I$ are permutation matrices, and thus automorphisms of the graph G.

The construction of the automorphism group may be facilitated by taking advantage of some additional information. Multiplying both sides of equation (11) on the right by $\mu_{A^2}(A)$ gives

$$P\mu_{A^2}(A) = \mu_{A^2}(A). \tag{13}$$

So if P is an automorphism of G, it can only interchange nodes of G corresponding to identical rows of $\mu_{A^2}(A)$. Moreover, if u and v are similar nodes, then the rows of $\mu_{A^2}(A)$ corresponding to u and v must constitute a minimal pair of identical rows. This follows from the fact that if $\mu_{A^2}(A)$ has more than two rows identical to the same one, the matrix $\mu_{A^2}(A) \sum_{i=0}^{p-m-1} b_i A^i$ will have a principal submatrix of order > 2 consisting of all ones.

<u>Corollary 3a</u>. Under the hypotheses of Theorem 3, a necessary condition for two nodes of a graph G to be similar is that the corresponding rows of $\mu_{A^2}(A)$ constitute a minimal pair of identical rows. Hence, if the rows of $\mu_{A^2}(A)$ are pairwise distinct, $\Gamma(G)$ is trivial.

Figure 5 exhibits two identity graphs which respectively satisfy and fail to satisfy the condition of Corollary 3a.

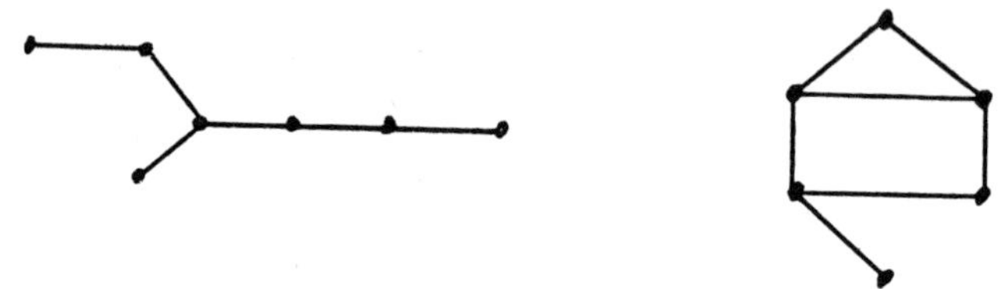

Rows of $\mu_{A^2}(A)$ are distinct

$\mu_{A^2}(A)$ has three minimal pairs of identical rows

Figure 5. Identity graphs

From Theorem 2, it follows that if the characteristic polynomial of the adjacency matrix of a graph is irreducible over GF(2), then its group is trivial. Of course, the polynomial might be irreducible over the integers and not over GF(2) -- for example the digraph of Figure 1 is irreducible over GF(2) but the graph is not (indeed the polynomial of a graph is always reducible over GF(2)). For a digraph with non-derogatory adjacency matrix, over GF(2), it remains to establish a general relationship between the factors of its characteristic polynomial and the cyclic subgroups of its automorphism group.

Appendix

The following table lists all connected graphs with p nodes ($2 \leq p \leq 5$) together with the characteristic and minimal polynomials of their respective adjacency matrices (over GF(2)).

	G	$\phi_{A(G)}(x)$	$\mu_{A(G)}(x)$
p=2		$(x+1)^2$	$(x+1)^2$
p=3		x^3	x^3
		$x(x+1)^2$	$x(x+1)$
p=4		$(x^2+x+1)^2$	$(x^2+x+1)^2$
		$x^2(x+1)^2$	$x(x+1)^2$
		x^4	x^2
		$(x+1)^4$	$(x+1)^4$
	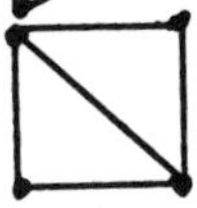	$x^2(x+1)^2$	$x(x+1)^2$
	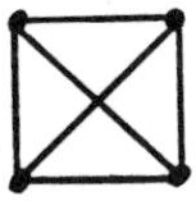	$(x+1)^4$	$(x+1)^2$

	G	$\phi_{A(G)}(x)$	$\mu_{A(G)}(x)$
p=5		$x(x+1)^4$	$x(x+1)^4$
	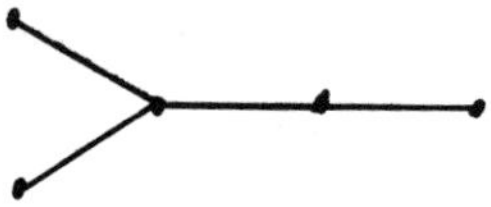	x^5	x^5
		x^5	x^3
	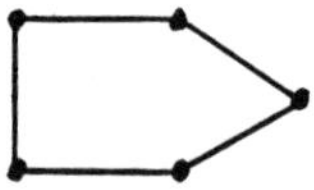	$x(x^2+x+1)^2$	$x(x^2+x+1)$
	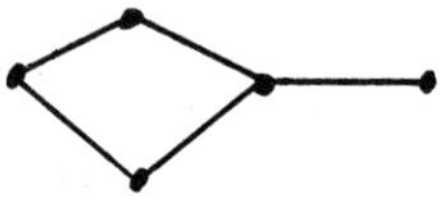	$x^3(x+1)^2$	$x^3(x+1)^2$
	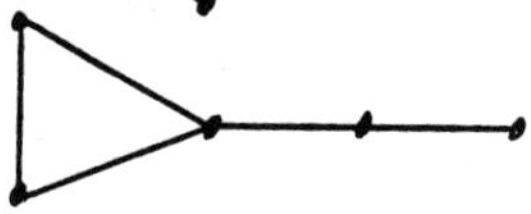	$x^3(x+1)^2$	$x^3(x+1)^2$
	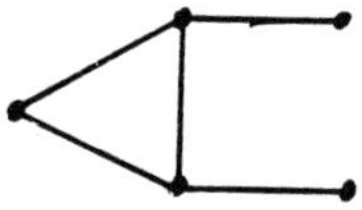	$x(x^2+x+1)^2$	$x(x^2+x+1)^2$
		$x^3(x+1)^2$	$x^3(x+1)$
	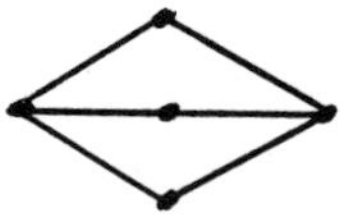	x^5	x^3
	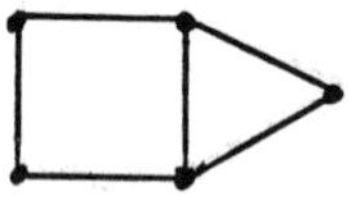	x^5	x^5
		$x(x+1)^4$	$x(x+1)^2$

p=5 (continued)

G	$A(G)^{(x)}$	$A(G)^{(x)}$
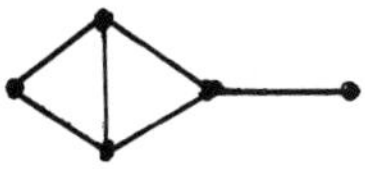	$x(x+1)^4$	$x(x+1)^4$
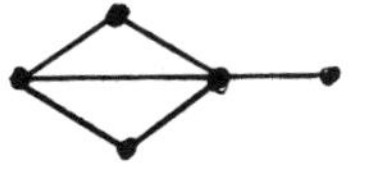	x^5	x^5
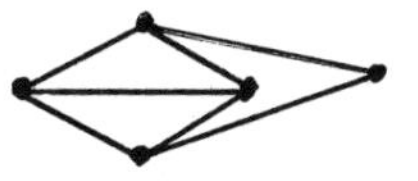	$x^3(x+1)^2$	$x^3(x+1)^2$
	$x(x^2+x+1)^2$	$x(x^2+x+1)^2$
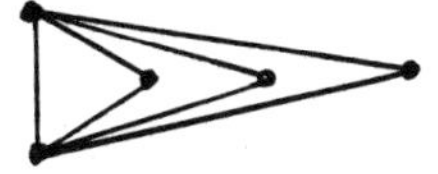	$x^3(x+1)^2$	$x(x+1)$
	$x^3(x+1)^2$	$x^3(x+1)$
	x^5	x^3
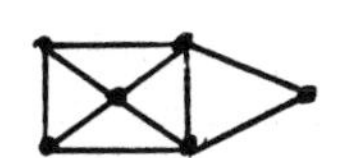	$x(x+1)^4$	$x(x+1)^2$
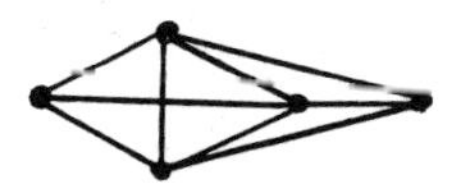	$x^3(x+1)^2$	$x^3(x+1)$
	$x(x+1)^4$	$x(x+1)$

REFERENCES

[1] C.-Y. Chao, On a theorem of Sabidussi. *Proc. Amer. Math. Soc.* 15 (1964), 291-292.

[2] C.-Y. Chao, On groups and graphs. *Trans. Amer. Math. Soc.* 118 (1965), 488-497.

[3] C.-Y. Chao, A note on the eigenvalues of a graph, *J. Combinatorial Theory* 10B (1971), 301-302.

[4] L. Collatz and A. Sinogowitz, Spektren endlicher Graphen. *Abh. Math. Sem. Univ. Hamburg* 21 (1957), 63-77.

[5] F. R. Gantmacher, *The Theory of Matrices*, Vol. I (English Translation) Chelsea, New York, 1959.

[6] F. Harary, *Graph Theory*, Addison-Wesley, Reading, 1969.

[7] F. Harary, R. Norman and D. Cartwright, *Structural Models*, Wiley, New York, 1965.

[8] A. Mowshowitz, The group of a graph whose adjacency matrix has all distinct eignevalues, *Proof Techniques in Graph Theory* (F. Harary, ed.) Academic Press, New York (1969), 109-110.

[9] A. Mowshowitz, The characteristic polynomial of a graph, *J. Combinatorial Theory*, to appear.

[10] A. Mowshowitz, Graphs, Groups and Matrices, Proceedings of the Canadian Mathematical Congress (June 1971), to appear.

[11] D. A. Suprunenko and R. I. Tyshkevich, *Commutative Matrices*, Academic Press, New York, 1968.

[12] H. Wielandt, *Finite Permutation Groups*, Academic Press, New York, 1964.

UNEXPLORED AND SEMI-EXPLORED TERRITORIES IN GRAPH THEORY

C. St. J.A. Nash-Williams

The ensuing remarks are the somewhat disorganized result of a request to outline what I believe to be some of the most promising directions for future research in graph theory. It seems convenient to subdivide this discussion under ten headings, although limitations of both space and my own knowledge compel some of these headings to receive only a passing mention.

The author apologizes for any errors, misrepresentations or deficiencies of style which may result from the fact that this written version of the lecture had to be drafted in extreme haste amid the complications of a transatlantic move.

1. Continuous Graph Theory

It seems likely that, to many theorems about discrete situations, there will correspond analogous theorems about continuous situations; and, so far as I know, virtually no attention has been paid to the probable existence of continuous counterparts of many of the theorems of graph theory, even though their discovery might require little more than a mechanical translation of discrete into continuous mathematics. The concepts of measure theory[1] seem to provide one obvious way of doing this. For example, according to one possible definition, a finite digraph is an ordered pair (V,E) where V is a finite set and E is a subset of $V \times V$. If the subset E of $V \times V$ is symmetric, this digraph may be called a graph. Letting 2^V denote the class of all subsets of V and N denote the set of non-negative integers, we automatically have available a useful function $\mu : 2^V \to N$ given by the rule that $\mu(S) = |S|$ for every $S \subset V$. A somewhat analogous continuous situation could now be described by defining a measure digraph to be an ordered pair $((V,B,\mu),E)$ where (V,B,μ) is a measure space and $\mu(V)$ is finite and $E \in B \times B$, i.e., E is a subset of $V \times V$ which is measurable in the cartesian product $(V \times V, B \times B, \mu \times \mu)$ of two copies of the measure

[1]In dealing with measure-theoretic concepts, I shall use the terminology and conventions of [1], with the additional constraint that, whenever a measure space (X,B,μ) is mentioned, it is understood that $X \in B$, so that B is a σ-algebra or Borel field.

space (V,B,μ). The measure digraph will be a measure graph if E is a symmetric subset of $V \times V$.

In a graph (V,E) or measure graph $((V,B,\mu),E)$, the elements of a subset S of V may be called scattered[2] if $(S \times S) \cap E = \emptyset$. Give positive integers n,k, Turán's Theorem [2, Theorem 13.4.1] determines the largest possible value of $|E|$ for graphs (V,E) such that $|V| = n$ and no k elements of V are scattered, and determines those graphs with the specified properties for which this largest value of $|E|$ is attained. Correspondingly, one might expect to be able to prove a theorem which determines, for any positive real number r and any positive integer k, the largest possible value of $(\mu \times \mu)(E)$ for measure graphs $((V,B,\mu),E)$ such that $\mu(V) = r$ and no k elements of V are scattered, and gives some information about which measure graphs with the specified properties achieve this largest value of $(\mu \times \mu)(E)$.

One could probably compile a very long list of other theorems in graph theory for which measure-theoretic analogues might appropriately be sought,

[2]The term "independent" is commonly used, but is is less favoured by the present author since it has other possible connotations important to combinatorics.

and I will here mention only three of the most obvious candidates. (I) The theorem[3] that the maximum number of edges in a matching in a bipartite graph equals the minimum cardinality of a set S of vertices such that every edge has at least one end in S [3, Theorem 10.2]. (II) The theorem which characterizes graphic sequences, i.e., sequences of numbers which can be realized by writing down in non-decreasing order the valencies of the vertices of some graph [3, Theorem 6.2]. (III) The theorem that, if (V,E) is a connected graph and f is a harmonic function from V into the set of real numbers [i.e., for each $v \in V$, f(v) is the average of f(w) over vertices w adjacent to v], then f is a constant function.

In constructing theorems along these lines, it might often be necessary to impose some reasonable restriction on the measure spaces which are used. Measure spaces of a reasonably continuous character, such as intervals on the real line or rectangular regions in Euclidean space of any dimension, would

---

[3]After giving this lecture, I was informed that Rado [36] proved a certain generalization of the above theorem which involves measure theory; but it does not appear to be precisely what I had in mind.

probably be entirely acceptable, but for many purposes one might have to impose restrictions which would exclude (or severely limit) measure spaces (V,B,μ) in which some measurable set E has the properties that $\mu(E) > 0$ and there is no measurable subset F of E with $0 < \mu(F) < \mu(E)$.

2. "Structural" Graph Theory

This very unsatisfactory heading is used for want of a better one to describe an area of graph theory embracing a rather wide variety of problems and results most of which nevertheless seem to exhibit a certain unity of character. Many of these results deal with graphs of a very general kind (as contrasted with, for example, results of the type mentioned in Section 5 below which are restricted to graphs in which most or all of the vertices have relatively high valencies) and describe how the shape of the graph affects the maximum possible size of something one wishes to find in it. For example, Menger's Theorem [3, Theorem 5.13] describes how the shape of a graph G affects the maximum number of disjoint paths by which two subsets of V(G) can be joined, and Tutte's Theorem [3, Theorem 9.4] which tells us how the shape of a graph determines whether or not it has a 1-factor (or "perfect matching") has been extended by Berge to a theorem [4, Chapter 18, Theorem 5] which describes how the shape of <u>any</u> graph determines the maximum of $|M|$ over all (perfect or imperfect) matchings M in the graph. We probably owe largely to the work of J. Edmonds the insight that graph-theoretic results of this type tend to be closely related to the duality theorem of linear programming, since, expressed loosely, the duality theorem of linear programming says that, in a wide class of situations, the largest possible size of something is equal to what is permitted by the worst of the constraints which tend to limit its size. Of course, the application of this duality theorem to situations in graph theory and combinatorics tends not to be straightforward, because combinatorial problems usually involve integer quantities whereas linear programming deals in the first instance with quantities which need not necessarily be integers; but sometimes the gulf between the integral situation of combinatorics and the non-integral situation of linear programming can be bridged. For example, Edmonds [5] has shown how to bridge this gulf with regard to the above mentioned matching theorems of Tutte and Berge by introducing the so-called "matching polyhedron", thus giving fundamental additional insight into these important theorems.

A problem which I have on various occasions suggested [6,7,8] is to try to characterize, for an arbitrary finite graph G and any prescribed disjoint subsets $X_1, X_2, \ldots, X_n$ of V(G), the maximum number of disjoint $X_1X_2\ldots X_n$-paths which can be found in G, where I call a path an <u>$X_1X_2\ldots X_n$-path</u> if it is an X_iX_j-path for some pair of distinct elements i,j of $\{1,2,\ldots,n\}$. A result of this nature might be interesting since it could unify three known theorems of "structural" graph theory, viz. Menger's Theorem (which is the solution to the special case n = 2 of the suggested problem), the theorem of Berge characterizing the maximum size of a matching in a graph (which is the solution to the special case of the suggested problem in which $X_1, \ldots, X_n$ are all the subsets of V(G) of cardinality 1) and a result of the present author [9] concerning a special case of the following problem: given a graph G and a function $f : V(G) \to N$ when can we find a set of edge-disjoint paths of positive length in G such that exactly f(v) of those paths end at each vertex v? (Embedding this problem in the original one about $X_1X_2...X_n$-paths involves constructing one graph from another, and we omit details.) I am indebted to J. Edmonds for pointing out that the special case of the $X_1X_2...X_n$-paths problem in which $X_1,...,X_n$ are some but not necessarily all of the one-element subsets of V(G) should also probably be regarded as being essentially solved, since it is equivalent to the following question whose treatment seems to be within the scope of existing techniques: if we form a graph G' from G by adding a loop at each vertex in $V(G) - (X_1 \cup ... \cup X_n)$ and define a function $h : V(G) \to N$ by letting $h(v) = 1$ for $v \in X_1 \cup ... \cup X_n$ and $h(v) = 2$ for $v \in V(G) - (X_1 \cup ... \cup X_n)$, what is the maximum possible number of edges in a spanning subgraph S of G' in which no vertex v has S-valency exceeding h(v)?

The preceding paragraph contains a suggestion for unifying some results in "structural" graph theory, and I can well envisage that this process of unification might be pursued much farther by integrating the sort of ideas outlined in the preceding paragraph with work of investigators such as J. Edmonds and T. A. Jenkyns, involving elements from graph theory, matroid theory and linear programming, to give something of which many of the main existing results of "structural" graph theory and matroid theory might be seen as special consequences. Combinatorics has sometimes been criticized as consisting of merely a number of isolated and unrelated problems with no central body of unifying theory. Whether or not this really constitutes a "criticism", any lack of central theory may be no more than a phenomenon characteristic of branches of mathematics in their infancy, and we may be approaching the time when there will be a unifying theory which embraces a very large part of "structural" graph theory.

An interesting unsolved problem in "structural" graph theory is provided by a conjecture concerning disjoint directed cutsets in digraphs which seems to have been thought of independently by D. H. Younger and N. G. Robertson. If G is a graph or digraph and $X \subset V(G)$, let $\overline{X}$ denote $V(G) - X$ and $X\delta$ denote the set of those edges of G which join vertices in X to vertices in $\overline{X}$. In the case in which G is a digraph, we call $X\delta$ a <u>directed cutset</u> or <u>dicutset</u> if either every edge in $X\delta$ is directed from its endvertex in X to its endvertex in $\overline{X}$ or every edge in $X\delta$ is directed from its endvertex in $\overline{X}$ to its endvertex in X. It is conjectured that for every finite digraph G there exists a non-negative integer n such that (i) G has n disjoint dicutsets and (ii) E(G) has a subset L such that $|L| = n$ and $L \cap X\delta$ is non-empty for every dicutset $X\delta$ of G, or, in other words, "the maximum number of disjoint dicutsets in a finite digraph equals the minimum cardinality of a set of edges which meets all dicutsets". Interesting partial results concerning this conjecture have been obtained by Younger [10].

Where problems in structural graph theory seem to be too difficult to solve (as in the case of the problem of characterizing graphs which have Hamiltonian circuits), it may be possible to formulate an easier, but presumably somewhat related, problem which _can_ be solved by replacing the integer quantities arising in the original problem by "fractional" quantities, i.e., quantities which are not restricted to integer values. To illustrate the idea, let us first note that the usual definition of a _Hamiltonian circuit_ of G is equivalent to defining it to be a spanning subgraph H of G such that (i) $|E(H) \cap X\delta| \geq 2$ for every non-empty proper subset X of V(G) and (ii) $|E(H)| = |V(G)|$. Of course, if H is a spanning subgraph of G, then any given edge of G is either in H or outside H: no intermediate possibility is allowed. However, if we were to relax this rather narrow-minded rule and allow an edge to be "partly in H" (e.g. two-thirds in H and one-third out), then a "spanning subgraph" H of the Petersen graph [3, Fig. 9.6] which includes exactly two-thirds of every edge would satisfy (i) and (ii) and hence be a "Hamiltonian circuit"; and thus we may say that the Petersen graph has a "fractional Hamiltonian circuit" in the sense of the following more precise definition. Call a function $f: E(G) \to R$ (where R is the set of real numbers) <u>pre-Hamiltonian</u> if $0 \leq f(\lambda) \leq 1$ for every $\lambda \in E(G)$ and $f \cdot X\delta \geq 2$ for every non-empty proper subset X of V(G), where the notation $f \cdot S$ means $\Sigma_{\lambda \varepsilon S} f(\lambda)$. It is easily seen that $f \cdot E(G) \geq |E(G)|$ for every pre-Hamiltonian function $f : E(G) \to R$, and we call a pre-Hamiltonian function $f : E(G) \to R$ a <u>fractional Hamiltonian circuit</u> of G if $f . E(G) = |V(G)|$. (Intuitively, $f(\lambda)$ tells us what fraction of the edge λ is in the fractional Hamiltonian circuit f.) Although it seems very difficult to characterize graphs which have Hamiltonian circuits, it might be realistic to look for some reasonable characterization of graphs with fractional Hamiltonian circuits, or, more generally, for some reasonable characterization in terms of the shape of a graph G of the minimum (h(G), say) of $f \cdot E(G)$ over all pre-Hamiltonian functions $f : E(G) \to R$. Since the latter question is a linear programming problem, presumably some kind of answer to it is furnished, at least in principle, by the duality theorem of linear programming; and perhaps this problem (which initially occurred to me many years ago before I was acquainted with linear programming) should really be formulated in the form: analyze the characterization of h(G) provided by the duality theorem of linear programming to ascertain whether it is, or can be translated into, something describable in a fairly simple and intuitive manner in terms of the shape of G.

The notion of "fractional Hamiltonian circuit" is taken up at greater length in the last section of the contribution to this volume by V. Chvátal, with certain modifications designed to strengthen the relationship between fractional Hamiltonian circuits and "integral Hamiltonian circuits", i.e.,

Hamiltonian circuits in the usual sense.

The reader will very probably be able to think of other fractional forms of known problems in graph theory. It might be added that this sort of language gives one way of indicating something of the spirit of Edmonds' work on the matching polyhedron. It is not hard to think of a definition of a "fractional matching", but if the most naive definition is selected then the resulting concept seems to be insufficiently closely related to ordinary matchings to provide really far-reaching insight. Edmonds' idea was essentially to select a less obvious definition of "fractional matchings" and then prove the existence of a much closer relationship between fractional matchings (in his sense) and matchings, viz. that the set of fractional matchings in a graph is, in a natural sense, the convex hull of the set of matchings in that graph.

Interesting and difficult results have in the last few years been obtained by Vizing [11, Theorem 14.4.1] and Gupta [12] concerning questions about how far it is true that the edges of a prescribed graph can be colored in a prescribed number of colors so that, for each vertex of the graph, a reasonably wide range of colors are used on the edges incident with that vertex. In this connection I repeat the following problem first formulated (so far as I know) in [8]. It is an easy exercise to prove that, if r is a positive integer, then a necessary condition for a regular r-valent graph to be r-edge-colorable is that

$$|X\delta| \geq r \text{ for every odd subset } X \text{ of } V(G)$$

(an <u>odd</u> set being a set of odd cardinality). It is also easily seen that (1) is a sufficient condition for r-edge-colorability when r = 2 but not (as the Petersen graph shows) when r = 3. Is the condition sufficient when r = 4? It is well known [3, Theorem 12.12] that a satisfactory characterization of 3-edge-colorable regular 3-valent graphs would settle the Four Color Problem, from which we might conclude that such a characterization is probably very hard to find; but it is at least conceivable that 4-edge-colorable regular 4-valent graphs might turn out to be easier to characterize because four edges at each vertex give us more room for maneuver than three and/or because graphs whose vertices have even valencies sometimes tend to be more tractable than more general graphs.

Finally, it may be worth calling renewed attention to the problem [13, problem 28] of characterizing those connected graphs whose vertices can be arranged in an order such that all distances between pairs of successive vertices in the ordering are at most 2. A closely related problem is to characterize those connected graphs whose vertices can be arranged in a cyclic order such that all distances between pairs of successive vertices are at most 2. Now that Fleischner [14,15] has overcome the difficult hurdle of proving that all nonseparable graphs admit the latter type of vertex-ordering, I think it fairly probable that the foregoing problem may be completely settled during the next few years.

3. Infinite Graph Theory

Since many combinatorial mathematicians tend to shun infinite structures, there is probably a wealth of manageable research problems concerning infinite graphs which have been neglected simply because relatively few mathematicians work in the area. D. König, who wrote the classical *Theorie der endlichen und unendlichen Graphen*, had a special predilection for infinite graphs, and the present author shares something of the feeling that they are particularly fascinating objects. Whilst the extension of some parts of finite graph theory to infinite graphs may be a straightforward and perhaps even tedious process, there are many other cases where handling infinite graphs calls for new ideas or techniques and poses a really interesting challenge. An example of the latter situation is the proof [18] of a counterpart for infinite graphs of Pósa's theorem on Hamiltonian circuits [3, Theorem 7.3], which involved methods almost entirely different from those hitherto used in proving Pósa's theorem and related results in finite graph theory, although I suspect that, in this and perhaps other instances, some useful feedback to finite graph theory might be obtained by applying to finite graphs the methods developed initially for handling infinite graphs.

Having surveyed results and problems in infinite graph theory at some length in [43], I shall comment only briefly on the subject here. As an example of a problem which might not be inordinately difficult to settle and may have remained unsolved mainly because nobody (amongst the few existing infinite graph-theorists) has found time to study it seriously, I might remind the reader of the following question of Erdös [13, problem 8]. Is it true that, if A and B are disjoint subsets of the set of vertices of an infinite graph G then there exists a subset S of V(G) which separates A from B (in the sense that every AB-path includes a vertex belonging to S) and a set P of disjoint AB-paths such that each element of S is on one of these paths and each path in P includes exactly one element of S? The answer would by Menger's Theorem be affirmative if "an infinite graph" were replaced by "a finite graph", but for infinite graphs the suggested assertion is stronger than the usual formulation of Menger's Theorem.

A _side_ of a graph G is a subset X of V(G) such that $X\delta = E(G)$. This terminology is motivated by a diagram such as Figure 1,

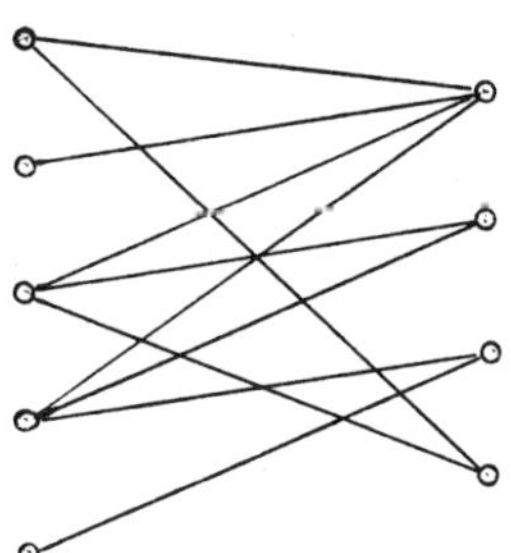

Figure 1

in which the vertices on the left side of the diagram constitute a side of the graph and those on the right side of the diagram also constitute a side of the graph. A graph is bipartite if it has a side. (It is easily shown that a nonempty connected bipartite graph has two and only two sides, which are complements of each other in the set of vertices of the graph.) It is well known [3, Chapters 5 and 10] that Menger's Theorem is closely related to the theory of matchings in bipartite graphs. If X is a side of an infinite bipartite graph G, one might ask when there exists a matching M in G such that each element of X is incident with an element of M, but questions of this type seem hitherto to have been answered only for the case in which X includes at most finitely many vertices of infinite valency [2, Theorem 7.3.3; 21, Theorem 4.3.1; 22, 23]: these results are often expressed in the language of systems of distinct representatives (also known as "transversals") of families of sets. I have recently made some suggestions (at present too ill-organized and perhaps a little too involved to be described in any detail here) as to how I think this theory might be extended to more general infinite bipartite graphs: discussions with

R. Rado have helped considerably in the formulation of these ideas. Should a development of the theory on these lines in fact prove possible, the same ideas might also yield an analogue for not necessarily locally finite graphs of Tutte's theorem characterizing finite graphs which have 1-factors: hitherto, this theorem has only been extended [24] to those infinite graphs which are <u>locally finite</u> (which means that the valencies of their vertices are all finite). Also, it seems not unreasonable to hope that approximately the same ideas might yield infinite versions, more general than those at present known, of known matroid-theoretic results [25,26,27,29,31] and their graph-theoretic consequences [29,32,33,34].

Finding a counterpart for infinite graphs of the result of Fleischner quoted at the end of Section 2 seems to be another natural project which might well be accomplished in the near future if anyone is sufficiently interested. Let us say that a denumerable graph G has a <u>one-way infinite k-ordering</u> if its vertices can be arranged in a one-way infinite sequence $\xi_1, \xi_2, \xi_3, \ldots$ such that the distance $d(\xi_i, \xi_{i+1})$ is less than or equal to k for each positive integer i, and let us say that G has a <u>two-way infinite k-ordering</u> if its vertices can be arranged in a two-way infinite sequence $\ldots, \xi_{-2}, \xi_{-1}, \xi_0, \xi_1, \xi_2, \ldots$ such that $d(\xi_i, \xi_{i+1}) \leq k$ for every integer i. In each case, it is understood that each vertex of the graph is to appear just once in the sequence. It is not true that every non-separable denumerable graph has a 2-ordering of one of these two types: the denumerable graph in Figure 2 provides a counterexample. Roughly speaking, the trouble is that this graph has "three

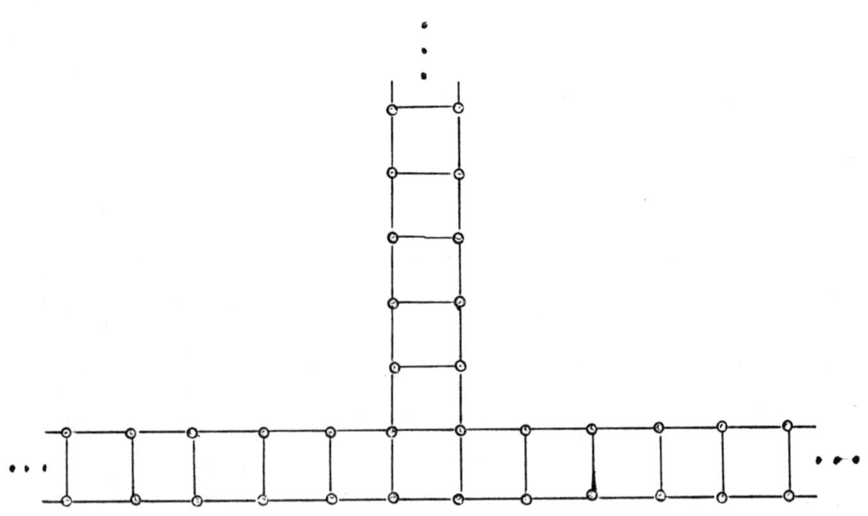

Figure 2

infinite wings branching out of a finite centre", and it is a matter of experience that graphs with "too many infinite wings branching out of a finite centre" must often be excluded in considering problems of this type: see, for example, the discussion in [43, Section 8] or [18, Section 1]. The idea of a graph G having "at least r infinite wings branching out of a finite centre" can be formalized by calling G r-divisible if there is a finite subset X of V(G) such that G - X has at least r infinite components. It is an easy exercise to prove [18, Section 1] that a graph which has one-way infinite Hamiltonian path must be 2-indivisible and a graph which has a two-way infinite Hamiltonian path must be 3-indivisible. Consequently, if G has a one-way infinite k-ordering, then G^k (the graph with the same vertices as G such that two vertices are adjacent in G^k if and only if they are distinct and the distance between them in G is $\leq$ k) must be 2-indivisible, and if G is a two-way infinite k-ordering then G^k must be 3-indivisible. Thus reasonable questions would be: (i) if G is a nonseparable denumerable graph such that G^2 is 2-indivisible, has G necessarily a one-way infinite 2-ordering? (ii) If G is a nonseparable denumerable graph such that G^2 is 3-indivisible, has G necessarily a two-way infinite 2-ordering? (iii) In the event of negative answers to (i) or (ii) being obtained, would these answers become affirmative if the conditions "G^2 is 2-indivisible" and "G^2 is 3-indivisible" were replaced by the stronger conditions "G is 2-indivisible" and "G is 3-indivisible" respectively? We note that similar questions concerning k-orderings of denumerable graphs where $k \geq 3$ have been studied by Sekanina [19,20].

The reader is referred to [35, pages 547-548] for another problem about Hamiltonian paths in infinite graphs.

4. Amusing Graph Theory

Interesting and apparently very difficult problems arise in connection with questions like: Given a positive integer r, does there exist a graph G which has no complete subgraph with 4 vertices and yet has the property that, no matter how we color the edges of G with r colors, there will be some complete subgraph of G with 3 vertices whose edges all receive the same color? For an interesting survey of this topic, I refer the reader to [37].

Graph-theoretic questions which may be of relatively little theoretical importance but can reasonably be classified as "amusing" arise in connection with chessboard problems such as questions about knight's tours etc. The positions on an m-by-n chessboard might be identified with the vertices of the graph $P_m \times P_n$, where P_m denotes a path with m vertices and the <u>product</u> $G \times H$ of two graphs G,H is a graph such that $V(G \times H) = V(G) \times V(H)$, $E(G \times H) = (V(G) \times E(H)) \quad (E(G) \times V(H))$, an element (ξ,μ) of $V(G) \times E(H)$ joins the vertices (ξ,η_1), (ξ,η_2) in $G \times H$ where η_1,η_2 are the vertices joined by μ in H, and an element (λ,η) of $E(G) \times V(H)$ joins the vertices $(\xi_1,\eta),(\xi_2,\eta)$ in $G \times H$ where ξ_1,ξ_2 are the vertices joined by λ in G. It is an easy exercise to see that $P_m \times P_n$ has a Hamiltonian circuit if and only if mn is even, $m \geq 2$ and $n \geq 2$. One might therefore ask more generally for a characterization of those pairs T_1,T_2 of finite trees for which $T_1 \times T_2$ has a Hamiltonian circuit. This is perhaps a slightly artificial and not outstandingly important problem; but it could pose an amusing challenge to mathematical ingenuity. The same question can be asked about Hamiltonian paths in denumerable products of trees and about products of three or more trees (where $T_1 \times T_2 \times T_3$ means $(T_1 \times T_2) \times T_3$, etc.). A little work on such questions has been done in [39],[40] and [41], but they remain largely unanswered.

I know of no complete determination of those rectangular chessboards $P_m \times P_n$ which admit knight's tours (i.e. sequences of moves whereby a knight occupies each position on the chessboard exactly once, possibly with the additional requirement that it must end up by returning to its starting-point), although a sufficiently diligent search of the literature of mathematics, mathematical amusements and chess might well discover a known answer. Even this type of question could be generalized to products of trees by, for example, defining a <u>knight's move</u> in $T_1 \times T_2$ to be a move from a vertex (ξ,η) to a vertex (ξ',η') such that one of the distances $d_{T_1}(\xi,\xi'), d_{T_2}(\eta,\eta')$ is 1 and the other is 2. Concerning knight's tours and related things in unbounded n-dimensional infinite "chessboards", I refer the reader to [44] and [43, Section 8].

5. Pósa-type Graph Theory

All graphs mentioned in Section 5 are understood to be finite and without loops or multiple edges.

Since a complete characterization of graphs with Hamiltonian circuits seems to be so far over the horizon, much research in Hamiltonian circuit theory has turned to the less ambitious project of finding interesting sufficient conditions for a graph to have a Hamiltonian circuit, and some of these conditions make considerable use of the <u>valency sequence</u> of the graph, i.e., the sequence obtained by arranging the valencies of its vertices in nondecreasing order. For example, let a sequence $a_1,\ldots,a_n$ of $n(\geq 3)$ integers be called a <u>Pósa sequence</u> if it satisfies the following conditions:

(i) $a_i \geq i + 1$ for every positive integer i less than $(n/2) - 1$,

(ii) $a_i \geq n/2$ for every integer i such that $n/2 \leq i \leq n$,

(iii) if n is odd, $a_{(n-1)/2} \geq (n - 1)/2$.

Pósa's theorem [3, Theorem 7.3] says that if the valency sequence of a graph is a Pósa sequence then the graph has a Hamiltonian circuit. In general, one might define "Pósa-type graph theory" as including, at least, all those results or conjectures which say that certain information about the valency sequence of a graph (possibly combined with additional information such as the hypothesis that the graph is nonseparable) ensures that it must have a Hamiltonian circuit, or, more generally, that it must contain a circuit of at least a certain length, and all results or conjectures of a similar type about digraphs. One might have the misgiving that results of this type seem a much inferior substitute for being able to find <u>necessary and sufficient</u> conditions for a graph to have a Hamiltonian circuit. On the other hand, it seems to be a characteristic of "Pósa-type graph theory" that there are many directions in which the available methods seem capable of pushing the existing frontiers of knowledge a little farther, provided that one invests a little time and effort in working out the details. This fact may possibly be a sign that, in the ultimate order of things, this area of graph theory is one that mathematicians were "meant" to investigate.

Having surveyed much of this subject area elsewhere [45,46,47], I shall mention only one specific question here. By a <u>digraph</u> I here mean a finite loopless digraph in which no two edges with the same tail (or initial vertex) may also have the same head (or final vertex), although we do permit a digraph to have edges λ,μ such that the tail of λ is the head of μ and the head of λ is the tail of μ. The <u>outvalency sequence</u> of a digraph is obtained by arranging the outvalencies of its vertices as a nondecreasing sequence, and its <u>invalency sequence</u> is similarly defined. One of the hard unsolved problems in this area of graph theory seems to be to decide whether a digraph whose invalency sequence and outvalency sequence are both Pósa sequences must necessarily have a Hamiltonian directed circuit. In this connection, it occurs to me that it is not even obvious whether, if α,β are distinct vertices of a digraph D whose invalency sequence and outvalency sequence are Pósa sequences, there must necessarily exist a directed circuit in D which passes through both α and β. I suggest this as a possibly easier problem which there might be some hope of solving, and whose solution might cast some light on the harder problem.

Amongst many references of interest concerning "Pósa-type graph theory", I mention [48],[49],[50] and [51].

6. Well-quasi-ordering Graphs

Let $\mathcal{G}$ be a class of graphs. A binary relation $\prec$ on $\prec$ is called a <u>well-quasi-ordering</u> of $\mathcal{G}$ if it is reflexive and symmetric and, for every infinite sequence $G_1, G_2, G_3, \ldots$ of graphs belonging to $\mathcal{G}$, there exist positive integers i,j such that $i < j$ and $G_i \prec G_j$. In particular, let $\leq$ be the binary relation on $\mathcal{G}$ defined by the rule that, for any two graphs $G, H \in \mathcal{G}$, we write $G \leq H$ if and only if some subgraph of H is isomorphic to a subdivision of G. A <u>subdivision</u> of G means, to express it a little informally, a graph obtainable from G by inserting a finite number of vertices of valency 2 in the middle of each edge (see Figure 3): the number of vertices inserted may be different for different edges of G, and is permitted to be zero. Thus, intuitively, $G \leq H$ means that G can be embedded in H if we are prepared to map an edge of G into a path of any positive finite length in H.

Difficult problems arise about whether certain binary relations are well-quasi-orderings of certain classes of graphs, and in particular about whether $\leq$ is a well-quasi-ordering of $\mathcal{G}$ for certain classes of graphs $\mathcal{G}$. Perhaps the most obvious unsolved

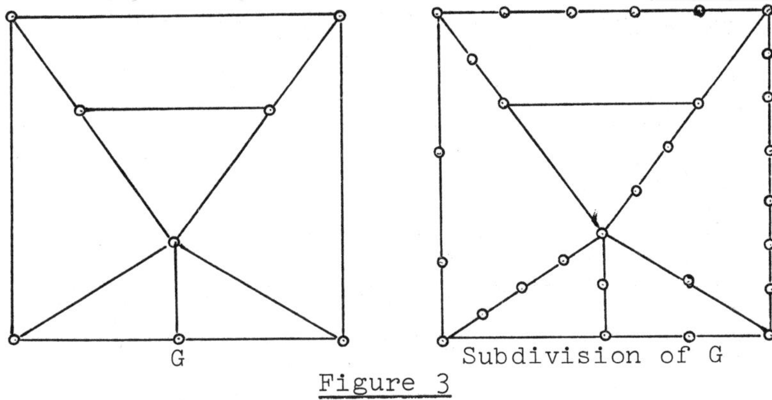

Figure 3

problem is to prove the conjecture of Vázsonyi that, if *G* is the class of graphs in which no vertices have valencies greater than 3, then $\leq$ is a well-quasi-ordering of *G* . Unfortunately this problem seems to be very difficult, and I hesitate to predict an early solution.

The interested reader may consult [53],[54] and [55] for some background to this subject area.

7. The Reconstruction Problem

Some discussion of this problem, which is recognized as one of the main unsolved problems of graph theory, may be found in [3, pages 12,13,41] and many other published references. If ξ is a vertex of a graph G, we call G - ξ (the graph obtained from G by removing ξ and its incident edges) a *vertex-deleted* subgraph of G. The reconstruction problem asks whether, in a certain sense, every finite graph with three or more vertices can be reconstructed from its vertex-deleted subgraphs, or, more precisely, whether it is true that a finite graph with three or more vertices is known *up to isomorphism* when we know its vertex-deleted subgraphs *up to isomorphism*. Still more precisely, let us say that two graphs G,H constitute a *reconstruction pair* if $|V(G)| = |V(H)| \geq 3$ and there exists a one-to-one function f from V(G) onto V(H) such that $G - \xi$ is isomorphic to $H - f(\xi)$ for every $\xi \in V(G)$: the problem is to determine whether there exists a reconstruction pair G,H with G and H finite for which G is not isomorphic to H.

Whilst this difficult problem may well be far from solution at the present time, there seems to be wide scope for research leading to partial results related to the reconstruction problem, such as (i) results proving the reconstruction conjecture for particular kinds of graphs, (ii) results saying that certain data about a graph G can be determined from knowing the vertex-deleted subgraphs of G up to isomorphism, (iii) study of problems similar to the reconstruction problem for structures other than, but perhaps bearing some resemblance to, graphs.

It is possible to find a reconstruction pair G,H such that G and H are infinite graphs and G is not isomorphic to H: a simple example, noted by J. Fisher, R. Graham and F. Harary [60], is obtained by taking G to be a tree in which every vertex has the same infinite valency and H to be a union of two disjoint isomorphic copies of this tree. This suggests the question: if G,H is a reconstruction pair such that both G and H are infinite trees, is it necessarily true that G is isomorphic to H? The corresponding question with both G and H infinite but locally finite forests has just been settled in the negative by F. Harary, R. Scott, and A. Schwenk [61].

8. Topological Graph Theory and the Four Color Conjecture

In addition to the Four Color Problem [57], this area of graph theory might be considered as including all questions involving graphs drawn in the plane or other surfaces and questions concerning the genus, thickness, crossing number etc.

[3, pages 116-123] of graphs: I have neither the time nor the competence to comment adequately on this important, interesting and vast area of graph theory which could by itself be the subject of several books. However, to avoid being completely vacuous, let me briefly raise some questions regarding one particular portion of this subject. A graph G will be called <u>4-connected</u> if $|V(G)| \geq 4$ and $G - X$ is connected for every subset X of V(G) such that $0 \leq |X| < 4$. Tutte has proved that every 4-connected planar graph has a Hamiltonian circuit [57, Theorem 5.2.2]. (i) Is there a 4-connected graph which can be drawn on the torus and has no Hamiltonian circuit? (ii) Has every 2-indivisible 4-connected denumerable planar graph a one-way infinite Hamiltonian path? (iii) Has every 3-indivisible 4-connected denumerable planar graph a two-way infinite Hamiltonian path?

9. Algebraic Graph Theory

Once again, this heading covers too wide a field to permit any even approximately adequate comment, since it could embrace all areas of interaction between algebra and graph theory. One of the most highly developed and interesting of these concerns the problem of determining the smallest possible number of vertices of a regular graph of prescribed girth and valency (the <u>girth</u> of a graph being the minimum of the lengths of the circuits in the graph). I tried to convey a few of the main ideas of this topic very briefly in [58, Section 6]. It has the curious feature that some of the main results, although purely combinatorial in character, seem in the present state of knowledge to be unobtainable without resorting to algebraic methods involving a consideration of eigenvalues of adjacency matrices of graphs.

Until some radically new ideas are discovered, it is possible that, in the problem of the smallest possible number of vertices of regular graphs of prescribed girth and valency, mathematicians have by now done most of what the available techniques will yield fairly easily; but nevertheless I will mention briefly one of the main outstanding problems. Is there a regular graph of valency 57 and girth 5 with as few as 3250 vertices? The construction of such a graph, if one exists, might be a challenge to experts in design theory; or the proof of its non-existence may call for a radical improvement in available algebraic, or other, techniques.

10. Other Graph Theory

It is not claimed that the previous nine sections (let alone the discussion under those headings) are exhaustive: for instance, I have not included "extremal graph theory" as one of these headings because there are others far more competent to delineate the main problems in this area. I believe that there may be some value in trying to coordinate and direct graph-theoretic thinking by indicating, in a *positive* sense, what may be some of the interesting directions in which it might go. It is not, however, the intention of this article to suggest, in a *negative* sense, that any research topics which do not receive such explicit mention are necessarily inferior. It would probably be a disservice to graph theory, or indeed to mathematics in general, to discourage new and original ideas by suggesting too dogmatically that research should always be confined to certain "approved" directions: combinatorial mathematicians should be particularly unready to take such a stand since they will remember that combinatorics itself was for a long period regarded, in many mathematical circles, as one of the unfashionable parts of mathematics. Of course, referees of articles, Ph.D. supervisors and others must, from time to time, make judgements discriminating between good and bad mathematical research; but at least let us evaluate each item on its own merits and not by setting up "fashions".

References

[1] P. R. Halmos, Measure Theory, Van Nostrand, Princeton, 1950.

[2] O. Ore, Theory of Graphs, American Mathematical Society, College Publ. Vol. 38, Providence, 1962.

[3] F. Harary, Graph Theory, Addison-Wesley, Reading, Mass., 1969.

[4] C. Berge, The Theory of Graphs and Its Applications, Dunod, Paris, 1958. (English Translation by A. Doig, Methuen, London, 1962.)

[5] J. Edmonds, Maximum matching and a polyhedron with 0,1-vertices, J. Res. Nat. Bur. Standards, 69B, 1965, 125-130.

[6] First Waterloo Combinatorics Conference, 1965 (unpublished).

[7] P. Erdos and G. Katona (editors), Theory of Graphs, Academic Press, New York, 1968.

[8] C. St. J.A. Nash-Williams, Possible directions in graph theory, [16], 191-200.

[9] C. St. J.A. Nash-Williams, Decomposition of finite graphs into open chains, Canad. J. Math. 13 (1961), 157-166.

[10] D. H. Younger, Maximum families of disjoint directed cut sets, [17], 329-333.

[11] V. G. Vizing, The chromatic class of a multigraph, Cybernetics (English translation of Kibernetika) 1 (1965), 32-41.

[12] R. P. Gupta, The cover index of a graph, Aequationes Math., to appear.

[13] M. Fiedler (editor), Theory of Graphs and its Applications, Czechoslovak Academy of Sciences, Prague, 1964.

[14] H. Fleischner, On spanning subgraphs of a connected bridgeless graph and their application to DT-graphs, J. Combinatorial Theory, Series B, to appear.

[15] H. Fleischner, The square of every two-connected graph is Hamiltonian, J. Combinatorial Theory, Series B, to appear.

[16] D.J.A. Welsh (editor), Combinatorial Mathematics and its Applications, Academic Press, London, 1971.

[17] W. T. Tutte (editor), Recent Advances in Combinatorics, Academic Press, New York, 1969.

[18] C. St. J.A. Nash-Williams, Hamiltonian lines in infinite graphs with few vertices of small valency, Aequationes Math., to appear.

[19] M. Sekanina, On an ordering of the set of vertices of a connected graph, Spisy Prirod, Fak. Univ. Brno, 412 (1960), 137-142.

[20] M. Sekanina, On an ordering of the set of vertices of a graph, Casopis Pest, Mat., 88 (1963), 265-282.

[21] L. Mirsky, Transversal Theory, Academic Press, London, 1971.

[22] J. Folkman, Transversals of infinite families with finitely many infinite members, RAND Corporation Memorandum RM-5676-PR, 1968.

[23] R. A. Brualdi and E. B. Scrimger, Exchange systems, matchings and transversals, J. Combinatorial Theory 5 (1968), 244-257.

[24] W. T. Tutte, The factorization of locally finite graphs, Canad. J. Math. 2 (1950), 44-49.

[25] J. Edmonds, Minimum partition of a matroid into independent subsets, J. Res. Nat. Bur. Standards 69B (1965), 67-72.

[26] J. Edmonds, Lehman's switching game and a theorem of Tutte and Nash-Williams, J. Res. Nat. Bur. Standards 69B (1965), 73-77.

[27] J. Edmonds, Submodular functions, matroids and certain polyhedra, [28], 69-87.

[28] R. K. Guy, H. Hanani, N. Sauer and J. Schönheim (editors), Combinatorial Structures and their Applications, Gordon and Breach, New York, 1970.

[29] C. St. J. A. Nash-Williams, An application of matroids to graph theory, [30], 263-265.

[30] C. Berge, Theory of Graphs, Dunod, Paris and Gordon and Breach, New York, 1967.

[31] J. S. Pym and H. Perfect, Submodular functions and independence structures, J. Math. Analysis Appl. 30 (1970), 1-31.
[32] W. T. Tutte, On the problem of decomposing a graph into n connected factors, J. London Math. Soc. 36 (1961), 221-230.
[33] C. St. J. A. Nash-Williams, Decomposition of finite graphs into forests, J. London Math. Soc. 39 (1964), 12.
[34] C. St. J. A. Nash-Williams, Edge-disjoint spanning trees of finite graphs, J. London Math. Soc. 36 (1961), 445-450.
[35] R. C. Bose, I. M. Chakravarti, T. A. Dowling, D. G. Kelly and K.J.C. Smith (editors), Proceedings of the Second Chapel Hill Conference on Combinatorial Mathematics and its Applications, 1970.
[36] R. Rado, A theorem on general measure functions, Proc. London Math. Soc. (2) 44 (1938), 61-91.
[37] R. L. Graham and J. H. Spencer, On small graphs with forced monochromatic triangles, [38], 137-141.
[38] M. Capobianco, J. B. Frechen and M. Krolik (editors), Recent Trends in Graph Theory, Springer-Verlag, Berlin, 1971.
[39] W. H. Cunningham, The circumference of a class of product graphs, [42], 203-224.
[40] C. St. J. A. Nash-Williams, Hamiltonian lines in products of infinite trees, J. London Math. Soc. 40 (1965), 37-40.
[41] K. A. Zaretskii, On the Hamiltonian cycle and the Hamiltonian path in the Cartesian product of two graphs, Cybernetics (English translation of Kibernetika), 2 (1966), 3-8.
[42] R. C. Mullin, K. B. Reid, D. P. Roselle and R.S.D. Thomas (editors), Proceedings of the Second Louisiana Conference on Combinatorics Graph Theory and Computing, 1971.
[43] C. St. J. A. Nash-Williams, Infinite graphs - a survey, J. Combinatorial Theory, 3 (1967), 286-301.
[44] C. St. J. A. Nash-Williams, Abelian groups, graphs and generalized knights, Proc. Cambridge Philos. Soc. 55 (1959), 232-238.
[45] C. St. J. A. Nash-Williams, Hamiltonian circuits in graphs and digraphs, [52], 237-243.

[46] C. St. J. A. Nash-Williams, Hamiltonian arcs and circuits, [38], 197-210.
[47] C. St. J. A. Nash-Williams, Valency sequences which force graphs to have Hamiltonian circuits (mimeographed).
[48] V. Chvátal, On Hamilton's ideals, J. Combinatorial Theory, to appear.
[49] G. A. Dirac, Some theorems on abstract graphs, Proc. London Math. Soc. 2 (1952), 69-81.
[50] J. A. Bondy, Large cycles in graphs, Discrete Math., to appear.
[51] L. Pósa, On the circuits of finite graphs, Magyar Tud. Akad. Mat. Kutató Int. Közl. 8 (1963), 355-361.
[52] G. Chartrand and S. F. Kapoor (editors), The Many Facets of Graph Theory, Springer-Verlag, Berlin, 1969.
[53] C. St. J. A. Nash-Williams, A survey of the theory of well-quasi-ordered sets, [28], 293-299.
[54] T. A. Jenkyns and C. St. J. A. Nash-Williams, Counterexamples in the theory of well-quasi-ordered sets, [56], 87-91.
[55] J. B. Kruskal, Well-quasi-ordering, the tree theorem and Vaszonyi's Conjecture, Trans. Amer. Math. Soc. 95 (1960), 210-225.
[56] F. Harary (editor), Proof Techniques in Graph Theory, Academic Press, New York, 1969.
[57] O. Ore, The Four-color Problem (Academic Press, New York and London, 1967 [Pure and Applied Mathematics, Vol. 27]).
[58] C. St. J. A. Nash-Williams, A survey of graph theory, [59], 383-444.
[59] R. C. Mullin, K. B. Reid and D. P. Roselle (editors), Proceedings of the Louisiana Conference on Combinatorics, Graph Theory and Computing, Louisiana State University, Baton Rouge, 1970.
[60] J. Fisher, R. Graham, and F. Harary, A simpler counterexample to the reconstruction conjecture for denumerable graphs, J. Combinatorial Theory, 12B (1972), 203-204.
[61] F. Harary, R. L. Scott, and A. J. Schwenk, On the reconstruction of countable forests, Publ. Inst. Math. (Beograd), to appear.

GRAPHICAL ENUMERATION METHODS

Edgar M. Palmer

Abstract

The development of selected graphical enumeration methods is sketched beginning with the work of Burnside, Redfield, and Pólya and continuing through the recent contributions of numerous combinatorialists. This provides the impetus for a suggested research program whose aim is not only directed toward the solution of specific counting problems but also the formulation of new methods and new directions.

1. Introduction.

Graphical enumeration is that part of the theory of graphs concerned with the determination by efficient means of the number of graphs of various kinds. The purpose of this article is to sketch past progress with selected topics in this area,

---

[1]Work supported in part by a grant from the National Science Foundation.

thereby indicating the directions in which the field seems to be headed. With a view toward continuing this progress, a research program is suggested which hopefully points in profitable and worthwhile new directions.

Graphical Enumeration is also the title of a forthcoming book [10] which contains the details of the methods discussed here as well as numerous applications. Definitions not given in this article may be found in the books [6,10].

2. Burnside's Lemma.

We shall trace the development of methods for counting graphs and digraphs. The number of labeled graphs of order p is easily seen to be $2^{\binom{p}{2}}$. Therefore to determine the number g_p of unlabeled graphs of order p, some method for stripping off the labels must be formulated. Therefore it is not too surprising that we begin with a group theoretic result called Burnside's Lemma [2], though it was undoubtedly known to Frobenius, Schur and others. If A is a permutation group with object set $X = \{1,2,\ldots,n\}$, then for each permutation α in A, $j_k(\alpha)$ denotes the number of cycles of length k in the disjoint cycle decomposition of α.

Burnside's Lemma. The number N(A) of orbits of the permutation group A is given by

$$N(A) = \frac{1}{|A|} \sum_{\alpha \varepsilon A} j_1(\alpha). \tag{1}$$

Usually the group A is some representation of the symmetric group which acts on labeled graphs. Then the problem is to find the number of labeled graphs left invariant by the permutations. The lemma forms the basis for countless enumeration theorems including those of Redfield [17] and Pólya [14]. In fact it was just formula (1) that Davis [3] used to count graphs and digraphs. Similarly, Robinson [19] counted even graphs, whose points all have even degree, by applying Burnside's Lemma and making use of a very clever scheme to determine the values of $j_1(\alpha)$ in (1).

To display the graph counting formula obtained from Burnside's Lemma, we require some definitions. We denote the partitions of p by vectors (j) = $(j_1, j_2, \ldots, j_p)$ where j_k is the number of parts equal to k. The g.c.d. and l.c.m. of r and t are denoted by (r,t) and [r,t] respectively. The next theorem [3,5] is obtained on applying Burnside's Lemma to an appropriate representation of the symmetric group S_p.

Theorem. The number g_p of graphs of order p is

$$(2) \qquad g_p = \frac{1}{p!} \sum_{(j)} \frac{p!}{\prod_k k^{j_k} j_k!} 2^{q(j)}$$

where the sum is over all partitions (j) of p and

$$(3) \quad q(j) = \sum_k [k/2] j_k + \sum_k k\binom{j_k}{2} + \sum_{r<t} (r,t) j_r j_t .$$

A slight modification of q(j) serves to yield d_p, the number of digraphs of order p (see [3,5]).

3. Redfield's Method.

Next we turn to the method of Redfield [17] which makes use of a polynomial Z(A) associated with a permutation group A. This polynomial in the variables $s_1, s_2, \ldots, s_n$ is called the cycle index because it keeps track of the disjoint cycles in all the permutations in A. By definition

$$(4) \qquad Z(A) = \frac{1}{|A|} \sum_{\alpha \in A} \prod s_k^{j_k(\alpha)} .$$

To display the variables we often write Z(A) = $Z(A; s_1, s_2, \ldots)$.

Redfield's enumeration theorem interprets the number of equivalence classes of m × n matrices obtained when an m-ary operation, denoted $\cap$ and called "cap", is performed on m cycle indexes of groups whose object sets each have n elements. For example, the next equation expresses the number N(A) of orbits of A as a cap-product:

$$(5) \qquad N(A) = Z(A) \cap Z(S_n)Z(S_1).$$

Burnside's Lemma follows quickly from (5).

To count graphs, Redfield required a certain representation of the symmetric group S_p now called the <u>pair group</u>, denoted $S_p^{(2)}$. For each permutation α in S_p, the corresponding permutation α' in $S_p^{(2)}$ acts on 2-subsets of X according to the rule

$$(6) \qquad \alpha'\{i,j\} = \{\alpha i, \alpha j\}.$$

Redfield devised an efficient algorithm for calculating the cycle index of a class of permutation groups to which $S_p^{(2)}$ belonged. Then his enumeration theorem could be applied to obtain the following formula for $g_{p,q}$, the number of (p,q)-graphs.

<u>Theorem</u>. The number $g_{p,q}$ of (p,q)-graphs is given by

$$(7) \qquad g_{p,q} = Z(S_p^{(2)}) \cap Z(S_q)Z(S_{k-q})$$

where $k = \binom{p}{2}$.

See [13] for a generalization of Redfield's Enumeration Theorem.

4. Pólya's Method.

Pólya's enumeration theorem [14] counts equivalence classes of functions from the set X to another set Y. With A acting on X, two functions f and h are equivalent if for some α in A and all x in X

$$f(x) = h(\alpha x). \tag{8}$$

Pólya's theorem not only provides a formula for the number of classes of functions in terms of the cycle index of A but also can be used to obtain the number of classes with certain prescribed properties. As reported in [5], Pólya found an explicit formula for $Z(S_p^{(2)})$, and with $A = S_p^{(2)}$, interpreted the equivalence classes of functions so that the polynomial

$$g_p(x) = \sum_{q=0}^{\binom{p}{2}} g_{p,q} x^q \tag{9}$$

which counts graphs of order p could be expressed in terms of $Z(S_p^{(2)})$.

Theorem. The polynomial $g_p(x)$ which enumerates graphs of order p is obtained from $Z(S_p^{(2)})$ by substituting $1+x^k$ for each variable s_k; symbolically:

(10) $$g_p(x) = Z(S_p^{(2)}, 1+x).$$

Slight modifications of (7) and (10) serve to count digraphs [5]. The permutation group required is also a representation of the symmetric group S_p. This representation is called the <u>reduced ordered pair group</u> and is denoted by $S_p^{[2]}$. For each permutation α in S_p, the corresponding permutation α' in $S_p^{[2]}$ acts on ordered pairs of distinct elements from the object set of S_p according to the rule

(11) $$\alpha'(i,j) = (\alpha i, \alpha j).$$

An explicit formula for the cycle index of $S_p^{[2]}$ can be found in many places, including [5,6,7,8,10].

<u>Theorem</u>. The polynomial $d_p(x)$ which enumerates digraphs of order p is obtained from $Z(S_p^{[2]})$ by substituting $1+x^k$ for each variable s_k; symbolically:

(12) $$d_p(x) = Z(S_p^{[2]}, 1+x).$$

Note that the total number d_p of digraphs of order p is obtained from (12) on setting $x = 1$.

5. <u>Power Group</u>.

deBruijn [1] generalized Pólya's Theorem by

determining a formula for the number of equivalence classes of functions from X to Y when two groups are involved: A with object set X and B with object set Y. Read [15] used this result to determine the numbers $\bar{g}_p$ and $\bar{d}_p$ of self-complementary graphs and digraphs:

Theorem. The numbers $\bar{g}_p$ and $\bar{d}_p$ are given by

$$\bar{g}_p = Z(S_p^{(2)};\ 0,2,0,2,\ldots), \tag{13}$$

$$\bar{d}_p = Z(S_p^{[2]};\ 0,2,0,2,\ldots). \tag{14}$$

A number of advantages are gained when deBruijn's theorem is reformulated by constructing the power group of A and B, see [7]. This group, denoted B^A, has as its object set the functions in Y^X. For each pair of permutations α in A and β in B there is a permutation in B^A denoted by $(\alpha;\beta)$ whose action on each function f in Y^X is defined by

$$((\alpha;\beta)f)(x) = \beta f(\alpha x) \tag{15}$$

for all x in X.

The next theorem, which is equivalent to deBruijn's, expresses the number of orbits of the power group B^A in terms of the cycle structure of the permutations in A and B.

Power Group Enumeration Theorem. The number $N(B^A)$ of orbits of the power group B^A is

$$(16) \qquad N(B^A) = \frac{1}{|B|} \sum_{\beta \in B} Z(A; c_1(\beta), c_2(\beta), \ldots),$$

where

$$(17) \qquad c_k(\beta) = \sum_{t|k} t j_t(\beta).$$

Now (16) can be verified by applying Burnside's Lemma to the power group. Furthermore the cycle index of the power group itself can be calculated [7] so that Pólya's theorem in turn can be applied to the power group. It was just such a procedure that served to determine the number d'_p and r'_p of self-converse digraphs and relations [8].

To count self-converse digraphs we restrict the power group B^A to the 1-1 functions in Y^X, and denote this restricted group by B^{A*}. Note that with E_2 denoting the identity group on two objects, the reduced ordered pair group $S_p^{[2]}$ and the restricted power group $S_p^{E_2*}$ are identical permutation groups. The restricted power group $S_p^{S_2*}$ contains in addition to the permutations of $S_p^{E_2*}$ an equal number of permutations which permit the coordinates of the pairs to be interchanged. To express the cycle index of $S_p^{S_2}$, we define a polynomial in the variables $s_1, s_2, \ldots$ as follows. For each permutation α in S_p, let

$$(18)\quad I(\alpha) = \prod_{k=1}^{p} s_{[2,k]}^{(2,k)k\binom{j_k}{2}} \cdot \prod_{1\leq r<t\leq p} s_{[2,r,t]}^{(2,[r,t])(r,t)j_r j_t}$$

$$\cdot \prod_{k \text{ odd}} s_{2k}^{((k-1)/2)j_k} \cdot \prod_{k \text{ even}} s_{k/2}^{\eta(k)2j_k} s_k^{(k-\eta(k))j_k}$$

where $\eta(k) = 0$ if $k/2$ is an odd integer and 1 otherwise. It is shown in [8] that $I(\alpha)$ is the contribution to $Z(S_p^{S_2^*})$ made when α permutes the ordered pairs as in (11) and is followed by switching coordinates. We denote the sum of these contributions by

$$(19)\qquad F(S_p^{S_2^*}) = \frac{1}{p!} \sum_{\alpha \varepsilon S_p} I(\alpha).$$

As observed in [8], the cycle index of $S_p^{S_2^*}$ can be expressed as the sum:

$$(20)\qquad Z(S_p^{S_2^*}) = \tfrac{1}{2}\{Z(S_p^{[2]}) + F(S_p^{S_2^*})\}.$$

Although $F(S_p^{S_2^*})$ is not the cycle index of a

permutation group, we will display the variables by writing $F(S_p^{S_2^*}; s_1, s_2, \ldots)$, and when s_k is replaced by $1+x^k$ we write $F(S_p^{S_2^*}, 1+x)$.

The counting theorem for self-converse digraphs depends on only that portion of $Z(S_p^{S_2^*})$ contributed by the permutations which interchange coordinates and takes the form:

Theorem. The polynomial $d_p'(x)$ which counts self-converse digraphs of order p is given by

$$d_p'(x) = F(S_p^{S_2^*}, 1+x). \tag{21}$$

Thus the number d_p' of self-converse digraphs is $F(S_p^{S_2^*}, 2)$.

We shall now collect our previous observations on the numbers of digraphs, self-complementary digraphs, and self-converse digraphs and use these to obtain the number of digraphs which have the same converse as complement as a new result. The complement of the digraph D is denoted by $\overline{D}$ and the converse by D'. It is the number $\overline{d}_p'$ of digraphs D of order p which have D' and $\overline{D}$ isomorphic that we seek. Note that this is not the same as the number of digraphs which are both self-complementary and self-converse, a much more difficult problem.

Just as self-complementary digraphs are enumerated by substituting alternating zeros and two's in the polynomial $Z(S_p^{[2]})$ in Formula (14), so can we obtain $\overline{d}'_p$ from $F(S_p^{S_2{}^*})$.

<u>Theorem</u>. The number $\overline{d}'_p$ of digraphs of order p for which the complement and the converse are isomorphic is given by

$$(22) \qquad \overline{d}'_p = F(S_p^{S_2{}^*};\ 0,2,0,2,\ldots).$$

We start a sketch of the proof by defining an equivalence relation on all labeled digraphs of order p. Two labeled digraphs D_1 and D_2 are <u>equivalent</u> if they are isomorphic, if D_1 and $\overline{D}_2$ are isomorphic or if D_1 and D'_2 are isomorphic. Formula (22) can be verified by expressing the number c_p of equivalence classes in two different ways.

The first way makes use of Burnside's Lemma (1) in a manner that is slightly different from the usual applications. We note that the two operations of taking the converse and the complement generate a group of transformations on the set of isomorphism classes of digraphs order p. This group has order 4 and the number of orbits it determines is c_p. To apply Burnside's Lemma (1) to this group we observe that the number of digraphs fixed by the identity transformation is d_p. The number fixed by complementation is $\overline{d}_p$ and the number fixed by taking the converse is d'_p. Finally, the number fixed by taking the complement and the converse is $\overline{d}'_p$. Therefore

$$c_p = \frac{1}{4}\{d_p + \overline{d}_p + d'_p + \overline{d}'_p\}. \tag{23}$$

The second way in which we express c_p involves the Power Group Enumeration Theorem. Consider the power group B^A with $B = E_2$ and $A = S_p^{E_2*}$. Since $S_p^{E_2*} = S_p^{[2]}$, the number of orbits $N(B^A)$ is simply d_p, the number of digraphs of order p, as in [8]. But if we take $B = S_2$ and keep $A = S_p^{E_2*}$, then $N(B^A)$ counts equivalence classes of digraphs for which two digraphs are equivalent whenever they are isomorphic or one is isomorphic to the complement of the other [7]. In [8], it was shown that with $B = E_2$ and $A = S_p^{S_2*}$, $N(B^A)$ again counts classes of digraphs. This time two digraphs are in the same class if they are isomorphic or one is isomorphic to the converse of the other. Thus with both $B = S_2$ and $A = S_p^{S_2*}$, $N(B^A)$ counts the classes when the digraphs in the same class are complementary <u>or</u> converse; that is

(24) $$c_p = N(B^A).$$

On applying (16) we have

(25) $$c_p = 1/2\{Z(S_p^{S_2^{*}}, 2) + Z(S_p^{S_2^{*}}; 0,2,0,2,\ldots)\}$$

Finally, c_p is eliminated from (23) and (25) and the resulting equation is solved for $\overline{d}'_p$. Formula (20) for $Z(S_p^{S_2^{*}})$ and Formulas (12),(14) and (21) which give $d_p, \overline{d}_p$ and d'_p respectively are then used to obtain the simplified form of (22).

The smallest values of $\overline{d}'_p$ were calculated using Formulas (18),(19) and (22) and are displayed in Table 1, along with the numbers g_p of graphs for comparison.

Table 1

p	1	2	3	4	5	6	7	8
g_p	1	2	4	11	34	156	1044	12346
$\overline{d}'_p$	1	1	2	5	20	88	632	8816

If α is the identity permutation, then from (18) we have $I(\alpha) = s_2^{\binom{p}{2}}$. Hence for all p

(26) $$\overline{d}'_p \geq 2^{\binom{p}{2}}/p!.$$

It appears from Table 1 that $\overline{d}'_p \leq g_p$. This suggests that $\overline{d}'_p$ is asymptotic to $2^{\binom{p}{2}}/p!$ as g_p is, see [11]. That this is indeed the case, can be verified using the techniques in [10, Chapter 9].

Corollary.

(27) $$\overline{d}'_p \sim 2^{\binom{p}{2}}/p!.$$

6. Cycle Index Sums.

None of the methods above serve to count blocks (nonseparable graphs), but a theorem of Redfield involving cycle index sums may have provided a hint which led to Robinson's successful attack on this difficult problem [18]. The main enumeration theorem in Robinson's paper establishes an identity for cycle index sums that has wide applicability. For example, if $\mathcal{H}$ is any set of graphs, denoted by $Z(\mathcal{H})$ the cycle index sum of the automorphism groups of all the graphs in $\mathcal{H}$. Let $\mathcal{G}$ and $\mathcal{C}$ denote the sets of all graphs and connected graphs respectively. Then Robinson's Theorem can be used to relate $Z(\mathcal{G})$ and $Z(\mathcal{C})$ as in the next equation where $s_k[Z(\mathcal{C})]$ denotes the sum obtained when the subscripts of all variables in $Z(\mathcal{C})$ are multiplied by k.

Theorem. The cycle index sums for graphs and connected graphs satisfy

$$(28) \qquad 1 + Z(\,G\,) = \exp \sum_{k=1}^{\infty} \frac{s_k}{k} [Z(\,C\,)].$$

Robinson provides an effective procedure for calculating $Z(\,G\,)$ so that (28) can be used to determine $Z(\,C\,)$, thus completing the first step in enumeration of blocks. He used this method not only to count blocks but also for the enumeration of connected graphs with no endpoints, homeomorphically irreducible graphs, cubic graphs, acyclic graphs and strong digraphs.

7. New Directions From Old.

In the preceding sections of this article, methods for solving selected graph counting problems were briefly sketched. The manner in which this past progress was made suggests to some extent the steps which may lead to continued progress in new directions. These steps seem to fall into two categories that are not necessarily unrelated:

Problems and Methods

(a) Problems

There is a large class of counting problems on which very little progress has been made. The methods available at the present time appear quite inadequate to cope with them. Although many of these problems can be stated in everyday language, they are deceptively difficult. For example, how many graphs of order p contain a triangle? How many have the property that each point is on a triangle? How many 2-dimensional simplicial complexes are there? How many digraphs are both self-complementary *and* self-converse?

For a comprehensive list of unsolved problems one may consult the survey article [9] or Chapter 10 of *Graphical Enumeration* [10]. It is to be hoped that many new interesting counting problems will also be formulated.

(b) *Methods*

Robinson's successful attack on the block counting problem led to a powerful general enumeration method that could be applied to numerous related problems. Similarly a solution to one of the problems involving triangles or simplicial complexes mentioned above could also lead to the solutions of others as well as a counting method.

Much of the progress of the past has come from the development of new methods which solve old problems. For example, all of the methods above can be used to count graphs. In fact, Read has shown [16] that many available enumeration formulas are more suitable for calculation when expressed in terms of Schur-functions; see also Foulkes [4]. But this interpretation requires a method for writing cycle indexes as sums of Schur-functions. Thus the reformulation and generalization of known methods may provide a fruitful new direction.

The methods of Read and Foulkes can be formulated in terms of group characters. Snapper [20] has also devised a counting method which makes use of the irreducible characters of the symmetric groups. His counting theorem expresses the number of (0,1)-matrices with prescribed row and column sums as an inner product of characters. The problem of specializing this theorem to symmetric matrices furnishes another interesting area for study.

We have seen how the pair group and the power group contributed to past advances. It is important therefore to seek new unary and binary operations on permutation groups <u>and</u> to attempt to establish associated cycle index formulas.

Another worthwhile direction involves asymptotics for graphs. Usually, these results (see [11], [12]) are based on the fact that the dominant term in Z(A;2,...,2) is contributed by the identity permutation. For example, Pólya found that $g_p \sim 2^{\binom{p}{2}}$. But this may not always be the case. Investigation of the asymptotic behavior of $Z(S_p^{(2)};\ 1,3,1,3,\ldots)$ may shed new light on this topic and lead to more refined asymptotic methods.

References

[1] N. G. deBruijn, Pólya's theory of counting, Applied Combinatorial Mathematics (E.F. Beckenbach, ed.) Wiley, New York, 1964, 144-184.

[2] W. Burnside, Theory of Groups of Finite Order (second edition) Cambridge University Press, Cambridge, 1911.

[3] R. L. Davis, The number of structures of finite relations, Proc. Amer. Math. Soc., 4 (1953) 486-495.

[4] H. O. Foulkes, On Redfield's range-correspondences, Canad. J. Math., 18 (1966) 1060-1071.

[5] F. Harary, The number of linear, directed, rooted, and connected graphs, Trans. Amer. Math. Soc., 78 (1955) 445-463.

[6] F. Harary, Graph Theory, Addison Wesley, Reading, 1969.

[7] F. Harary and E. M. Palmer, The power group enumeration theorem, J. Combinatorial Theory, 1 (1966) 157-173.

[8] F. Harary and E. M. Palmer, Enumeration of self-converse digraphs, Mathematika, 13 (1966) 151-157.

[9] F. Harary and E. M. Palmer, A survey of graphical enumeration problems, A Survey of Combinatorial Theory, (J. Srivastava et. al., eds.)

North-Holland, Amsterdam, 1972, to appear.
[10] F. Harary and E. M. Palmer, Graphical Enumeration, Academic Press, New York, 1973, to appear.
[11] W. Oberschelp, Kombinatorische Anzahlbestimmungen in Relationen, Math. Ann., 174 (1967) 53-58.
[12] E. M. Palmer, Asymptotic formulas for the number of self-complementary graphs and digraphs, Mathematika, 17 (1970) 85-90.
[13] E. M. Palmer and R. W. Robinson, The matrix group of two permutation groups, Bull. Amer. Math. Soc. 73 (1967) 204-207.
[14] G. Pólya, Kombinatorische Anzahlbestimmungen für Gruppen, Graphen und chemische Verbindungen, Acta Math., 68 (1937) 145-254.
[15] R. C. Read, On the number of self-complementary graphs and digraphs, J. London Math. Soc., 38 (1963) 99-104.
[16] R. C. Read, The use of S-functions in combinatory analysis, Canad J. Math., 20 (1968) 808-841.
[17] J. H. Redfield, The theory of group-reduced distributions, Amer. J. Math., 49 (1927) 433-455.
[18] R. W. Robinson, Enumeration of non-separable graphs, J. Combinatorial Theory, 9 (1970) 327-356.
[19] R. W. Robinson, Enumeration of euler graphs, Proof Techniques in Graph Theory, (F. Harary, ed.), Academic Press, New York, 1969, 147-153.
[20] E. Snapper, Group characters and nonnegative integral matrices, J. Algebra 19 (1971), 520-535.

ISOMORPHISMS BETWEEN HYPERGRAPHS

Richard Rado*

1. Notation, definitions and preliminary results.

Roman capital letters denote sets, and $|A|$ denotes the cardinal number of A. The relation $A \subset B$ denotes inclusion in the wide sense, and the relation $A \subset\subset B$ expresses the condition that A is finite and $A \subset B$. All sets are permitted to be empty, finite or infinite unless the contrary is stipulated. If $X \subset A$ and $f : A \to B$ then $f(X) = \{f(x) : x \in X\}$. We risk no confusion on account of the ambiguity in this last notation which exists whenever some element of A is at the same time a subset of A. If N is a set and A_i a set for each $i \in N$ then $\Gamma = (A_i : i \in N)$ denotes a _family of sets_ or, in another terminology, a _hypergraph_. The node set of Γ is $\bigcup(i \in N)A_i$ and its _edge set_ is $\{A_i : i \in N\}$. We call Γ an N-family. For the general theory of hypergraphs see [7].

*Research supported by a Canadian Commonwealth Research Fellowship.

For $M \subset N$ we put $A_M = \bigcup(i \in M)A_i$, and if $M \neq \phi$ then $A_{[M]} = \bigcap(i \in M)A_i$. Throughout this article, N denotes some fixed set; j,i range over the elements of N and M over the <u>non-empty</u> subsets of N.

As is well known, the relation $|A| = |B|$ means that there is a bijection $A \to B$. We will now extend this relation from a pair of sets to a pair of N-families.

<u>Definition</u>. The relation

(1) $$|A_i : i \in N| = |B_i : i \in N|$$

means that there is a bijection $f : A_N \to B_N$ such that, for each i, $f(A_i) = B_i$. Thus, (1) is true if $N = \phi$, and if $N = \{k\}$ then (1) is equivalent to $|A_k| = |B_k|$ in classical notation. Another terminology is to say that (1) expresses <u>strong isomorphism</u> between hypergraphs, or <u>isomorphism</u> between families. Instead of (1) we shall also use the notation

$$(A_i : i \in N) \cong (B_i : i \in N).$$

Clearly, (1) is an equivalence relation on the class of all N-families; following Bertrand Russell we may think of the symbol $|A_i : i \in N|$ as denoting the equivalence class to which the family $(A_i : i \in N)$ belongs, and we propose to call $|A_i : i \in N|$ a

<u>multi-cardinal</u>.

As illustration we quote the relations

$$|\{0\},\{1\}| = |\{3\},\{2\}| \neq |\{0\},\{0\}|;$$

$$|\{0,2\},\{0,1\},\{0,3\}| = |\{0,1\},\{0,2\},\{0,3\}|$$

$$\neq |\{0,1\},\{1,2\},\{2,3\}| \neq |\{1,2\},\{0,1\},\{2,3\}|.$$

We shall now extend the relation $|A| \leq |B|$ to multi-cardinals.

<u>Definition</u>. The relation

$$|A_i : i \in N| \leq |B_i : i \in N|$$

means that there is an injection $f : A_N \to B_N$ such that, for each i, $f(A_i) \subset B_i$. It is easy to verify that this relation between N-families is transitive. We remark that the obvious extension of the Schröder-Bernstein Theorem does not hold for N-families. For let $P = \{1,2,...\}$ and $Q = \{-1,-2,...\}$. Then

$$|P, P \cup Q| \leq |P \cup Q, P \cup Q|$$

as we may take as f the identity function. We also have

$$|P \cup Q, P \cup Q| \leq |P, P \cup Q|.$$

To see this take any injection of the denumerable set P Q into the denumerable set P. However, the relation

$$|P, P \cup Q| = |P \cup Q, P \cup Q|$$

is false. For, no bijection $g : P \cup Q \to P \cup Q$ satisfies $g(P) = P \cup Q$. Thus for $|N| \geq 2$, the inequality $\leq$ between N-families defines in the class of all N-families a partial pre-order which is not a partial order. However, if the A_i and B_i are finite, then this partial pre-order is, in fact, a partial order, by the following theorem.

<u>Theorem 1</u>. <u>Let</u> $|A_i : i \in N| \leq |B_i : i \in N|$ <u>and</u>, <u>for each</u> i, $|B_i| \leq |A_i| < \aleph_0$. <u>Then</u>
$|A_i : i \in N| = |B_i : i \in N|$.

<u>Proof</u>. There is an injection $f : A_N \to B_N$ satisfying $f(A_i) \subset B_i (i \in N)$. Then, for each i,
$|A_i| = |f(A_i)| \leq |B_i| \leq |A_i| < \aleph_0$. Hence $f(A_i)=B_i$ and the conclusion follows.

Our next object is to find conditions on ordinary cardinal numbers which will ensure that

(2) $$|A_i : i \in N| \leq |B_i : i \in N|,$$

and similarly for $|A_i : i \in N| = |B_i : i \in N|$.
For a future application we note that if (2) holds, then $|A_j \cup A_i| \leq |B_j \cup B_i|$ and, more generally,

(3) $$|\bigcup(k \in L)A_{[M_k]}| \leq |\bigcup(k \in L)B_{[M_k]}|$$

for all choices of L and M_k.

We recall the following well known definitions and propositions. The <u>boolean atoms</u>, briefly <u>atoms</u>, of the family $(A_i : i \in N)$ are the sets

$$\alpha_M = A_{[M]} - A_{N-M}$$

which are defined for all M. Clearly, α_M is the set of all elements of A_N which lie in A_i for all $i \in M$ and which do not lie in any A_i with $i \notin M$. The atoms of $(A_i : i \in N)$ will always be denoted by α_M, those of $(B_i : i \in N)$ by β_M, and so on.

<u>Theorem 2</u>. (i) $\alpha_M \cap \alpha_{M'} = \phi$ <u>for</u> $M \neq M'$.

(ii) <u>For each</u> i, $A_i = \bigcup(M \ni i)\alpha_M$.

<u>Proof of (i)</u>. We may assume that there is $i \in M-M'$. Then $\alpha_M \subset A_{[M]} \subset A_i$; $\alpha_{M'} = A_{[M']} - A_{N-M'} \subset A_N - A_i$.

<u>Proof of (ii)</u>. Let $i_o \in N$; $x \in A_{i_o}$; $M_x = \{j : A_j \ni x\}$. Then $i_o \in M_x$; $x \in A_{[M_x]} - A_{N-M_x} = \alpha_{M_x} \subset \bigcup(M \ni i_o)\alpha_M$ and hence $A_{i_o} \subset \bigcup(M \ni i_o)\alpha_M$. <u>Vice versa</u>, if $i_o \in M$ then $\alpha_M \in A_{i_o}$. Thus $\bigcup(M \ni i_o)\alpha_M \subset A_{i_o}$.

The atoms of a family can be prescribed arbitrarily subject of (i) above:

<u>Theorem 3</u>. <u>For each</u> M, <u>let</u> $\alpha(M)$ <u>be a set that</u> $\alpha(M) \cap \alpha(M') = \phi$ <u>for</u> $M \neq M'$. <u>Put, for each</u> i, $A_i = \bigcup(M \ni i)\alpha(M)$. <u>Then</u> $\alpha_M = \alpha(M)$ <u>for all</u> M.

Proof. Let M be fixed. Then $i \in M$ implies $\alpha(M) \subset A_i$. Hence $\alpha(M) \subset A_{[M]}$. On the other hand, $i \not\in M$ then $\alpha(M) \cap A_i = \phi$, so that $\alpha(M) \cap A_{N-M} = \phi$. Therefore $\alpha(M) \subset A_{[M]} - A_{N-M} = \alpha_M$. If now M is made to vary, then we have $A_N = \bigcup(M \subset N)\alpha(M) \subset \bigcup(M \subset N)\alpha_M = A_N$, and here both the set unions are disjoint. Hence $\alpha(M) = \alpha_M$ for all M.

The connection between atoms and multi-cardinals is given by

Theorem 4. The relation $|A_i : i \in N| = |B_i : i \in N|$ holds if and only if, for all M, $|\alpha_M| = |\beta_M|$.

Proof. If f is a bijection $A_N \to B_N$ such that, for all i, $f(A_i) = B_i$ then, for all M, $f(\alpha_M) = \beta_M$ and so $|\alpha_M| = |\beta_M|$.

Let $|\alpha_M| = |\beta_M|$ for all M. Then there are bijections $f_M : \alpha_M \to \beta_M$. By Theorem 1 (i), there is a function $f : A_N \to B_N$ whose restriction to α_M is f_M, for all M. Then f is bijective and satisfies, for each i.

$$f(A_i) = f(\bigcup(M \ni i)\alpha_M) = \bigcup(M \ni i)f(\alpha_M)$$
$$= \bigcup(M \ni i)\beta_M = B_i.$$

At the risk of being rather too obvious, I note that the relation $|A_i : i \in N| \leq |B_i : i \in N|$ does not imply that $|\alpha_M| \leq |\beta_M|$ for all M, though this latter inequality must certainly hold for some M. Consider the case $N = \{1,2\}; A_1=\{a\}; A_2=B_1=B_2=\{a,b\}$, where $a \neq b$. Then $|A_1,A_2| \leq |B_1,B_2|$ but $\alpha_{\{2\}} = \{b\}$ whereas $\beta_{\{2\}} = \phi$.

We can use Theorems 2-4 to obtain a method of coding the N-families, i.e., of associating with each N-family a code symbol whose variability range is easily described and which has the property that two N-families are isomorphic if and only if there code symbols agree.

Let us put $\underline{cod}(A_i : i \in N) = \{(M,|\alpha_M|) : M \subset N\}$. It is clear from Theorems 2-4 that two N-families are isomorphic if and only if their codes cod agree; also, given any cardinals τ_M there is an N-family whose code is $\{(M,\tau_M) : M \subset N\}$.

If N is finite, say $N = \{1,2,\ldots,n\} \neq \phi$, then instead of the code cod we can use the more convenient code cd which is defined as follows. Arrange all M in a sequence $M_1,M_2,\ldots,M_{2^n-1}$ in such a way that if $M = \{a_1,\ldots,a_r\}$; $M' = \{b_1,\ldots,b_s\}$; $r,s \geq 1$; $a_1 < \ldots < a_r$; $b_1 < \ldots < b_s$ then M precedes M' if either (i) $r < s$, or (ii) $r = s$ and $(a_1,\ldots,a_r)$ precedes $(b_1,\ldots,b_s)$ in the lexicographical ordering. Put

$$cd(A_1,\ldots,A_n) = (|\alpha M_1|, |\alpha M_2|, \ldots, |\alpha_{M_{2^n-1}}|).$$

Thus, if a,b,c are distinct objects then

$$cd(\{a,b\},\{a,c\},\{b\}) = (0,1,0,1,1,0,0).$$

Every coding opens the way to enumeration:

<u>Theorem 5</u>. Let $|N| = n > 0$.

(i) <u>The number of isomorphism types of</u> N-<u>families with</u> $|A_N| = a$ <u>is</u>

$$\binom{2^n - 2 + a}{a}.$$

(ii) <u>The number of isomorphism types of such families with</u> $|A_N| \leq b$ <u>is</u>

$$\binom{2^n - 1 + b}{b}.$$

<u>Proof of (i)</u>. Using the code <u>cd</u> we find that the required number equals the number of sequences $(a_1,\ldots,a_k)$ with $k = 2^n - 1$; $a_1,\ldots,a_k \in \{0,1,2,\ldots,a\}$; $a_1 + \ldots + a_k = a$. This equals the number of sequences $(a_1 + 1, a_1 + 2, \ldots, a_1 + \ldots + a_{k-1} + k-1)$ with the same restrictions, and of the latter there are exactly

$$\binom{a + k - 1}{k - 1}.$$

Proof of (ii).

$$\sum(0 \le a \le b)\binom{a+k-1}{a} = \binom{b+k}{b}.$$

To illustrate Theorem 5, we list representatives of the six different types of families (A_1, A_2) with $|A_1 \cup A_2| = 2$ $(a \neq b)$:

$(\phi,\{a,b\}),(\{a\},\{b\}),(\{a\},\{a,b\}),(\{a,b\},\phi),$
$(\{a,b\},\{a\}),(\{a,b\},\{a,b\}).$

After this digression on coding we return to order relations between N-families. It is useful to introduce one more relation of this kind, the <u>weak order</u> motivated by (3) above:

(4) $$|A_i : i \in N| \lesssim |B_i : i \in N|$$

means that whenever L is a set and, for each $k \in L$, M_k is arbitrary then

(5) $$|\bigcup(k \in L)A_{[M_k]}| \le |\bigcup(k \in L)B_{[M_k]}|.$$

Clearly, the relation (4) is transitive. If $A_j \cap A_i = \phi$ for $j \neq i$ then (4) is the same as

$$|B_M| \ge |A_M| \quad \text{for all } M.$$

If, in addition, $|A_i| = 1$ for all i, then this reduces to

(6) $$|B_M| \ge |M| \quad \text{for all } M,$$

a condition known as <u>Hall's condition</u> for the family $(B_i : i \in M)$. On the other hand, the relation

$$(7) \qquad |B_i : i \in N| \geq |\{i\} : i \in N|$$

means that the family $(B_i : i \in N)$ possesses a <u>system of distinct representatives</u>. To see this, assume first that (7) holds. Then there is an injection $f : N \to B_N$ such that $f(\{i\}) \subset B_i$, i.e., $f(i) \in B_i$. Then $f(j) \neq f(i)$ if $j \neq i$, and the elements $f(i)$ constitute a system of distinct representatives of the family $(B_i : i \in N)$. Vice versa, let $b_i (i \in N)$ be a system of distinct representatives of this family. Define f on N by putting $f(i) = b_i$ for $i \in N$. Then f is an injection $N \to B_N$ such that $f(\{i\}) \quad B_i$, and (7) follows.

Let us explore the relationship between the strong order $\leq$ and the weak order $\lesssim$.

<u>Theorem 5'</u>. <u>If</u> $|A_i : i \in N| \leq |B_i : i \in N|$ <u>then</u> $|A_i : i \in N| \lesssim |B_i : i \in N|$.

<u>Proof</u>. There is an injection $f : A_N \to B_N$ such that, for each i, $f(A_i) = B_i' \subset B_i$. Then, for every choice of L and of the M_k,

$$|\bigcup(k \in L) A_{[M_k]}| = |\bigcup(k \in L) B'_{[M_k]}| \leq |\bigcup(k \in L) B_{[M_k]}|.$$

The weak order is strictly weaker than the strong order. For let $N = \{1,2,3,...\}$; $A_i = \{a_i\}(i \varepsilon N)$; $B_1 = \{b_1, b_2, ...\}$; $B_i = \{b_{i-1}\}$ for $i > 1$. Here $a_1, a_2, ...$ are pairwise distinct and $b_1, b_2, ...$ are pairwise distinct. We have $|A_i : i \varepsilon N| \lesssim |B_i : i \varepsilon N|$. For, as the A_i are mutually disjoint, this is the same as $|A_M| \leq |B_M|$ for all M. Now, if $1 \varepsilon M$ then

$$|A_M| \leq |A_N| = |B_1| \leq |B_M|,$$

and if $1 \not\varepsilon M$ then $|A_M| = |M| = |B_M|$.

However, $|A_i : i \varepsilon N \not\lesssim |B_i : i \varepsilon N|$. For let f be a function such that $f : A_N \to B_N$ and $f(A_i) \subset B_i$ for all i. Then there is $k \geq 1$ such that $f(a_1) = b_k$. Then $f(a_{k+1}) \varepsilon f(A_{k+1}) \subset B_{k+1} = \{b_k\}$, so that $f(a_1) = f(a_{k+1}) = b_k$. Thus f is not injective, which makes the strong order relation between our two families impossible to hold. This is the well known counterexample to the obvious extension of Hall's theorem [1] to the case of infinite N.

There is a certain amount of overlap of the contents of the present section with [2]. However, the notation in [2] does not agree with the notation used here. If in [2] the measure of X is taken to be the cardinal number of X then

$|A_i : i \varepsilon N| \leq |B_i : i \varepsilon N|$ in [2] means $|A_{i''}| \lesssim |B_{i''}|$ here,
$|A_{i''}| = |B_{i''}|$ in [2] means $|A_{i''}| \lesssim |B_{i''}| \lesssim |A_{i''}|$ here,
$|A_{i''}| \equiv |B_{i''}|$ in [2] means $|A_{i''}| = |B_{i''}|$ here.

2. Hall's theorem and extensions.

Theorem 6 (P. Hall, [1]). If $|N| < \aleph_0$ and $|\{i\} : i \varepsilon N\}| \lesssim |B_i : i \varepsilon N|$ then $|\{i\} : i \varepsilon N| \leq |B_i : i \varepsilon N|$.

This means that for finite N Hall's condition (6) is necessary and sufficient for the existence of a system of distinct representatives. An easy extension is the following.

Theorem 7. Let $|N| < \aleph_0$; $|A_N| < \aleph_0$ and $A_j \cap A_i = \phi$ for $j \neq i$. Let $|A_i : i \varepsilon N| \lesssim |B_i : i \varepsilon N|$. Then $|A_i : i \varepsilon N| \leq |B_i : i \varepsilon N|$.

Proof. Put $B_{ix} = B_i$ for $i \varepsilon N$ and $x \varepsilon A_i$. We first show that

(8) $|\{(i,x)\} : i \varepsilon N; x \varepsilon A_i| \lesssim |B_{ix} : i \varepsilon N; x \varepsilon A_i|$.

We may assume that $A_i \neq \phi$ for all i. By definition of $\lesssim$ the relation (8) has the following meaning. Choose M and, for each $i \varepsilon M$, a non-empty set $A_i' \subset A_i$. Then

(9) $|\{(i,x) : i \varepsilon M; x \varepsilon A_i'\}| \leq |\bigcup(i \varepsilon M; x \varepsilon A_i')B_{ix}|$.

In fact, (9) is the same as $|A'_M| \leq |B_M|$ which is true by hypothesis. Having established (8), we can now apply Theorem 6. We obtain pairwise distinct elements b_{ix} such that $b_{ix} \varepsilon B_{ix}$ for $i \varepsilon N$ and $x \varepsilon A_i$. Define a function f on A_N by putting $f(x) = b_{ix}$ for $i \varepsilon N$ and $x \varepsilon A_i$. Then f is injective and $f(A_i) \subset B_i$ for all i. This completes the proof.

A less trivial extension is the following theorem.

Theorem 8 (M. Hall [3]). *Let* N *be arbitrary and* B_i *finite for all* i, *and* $|B_M| \geq |M|$ *for all finite* M. *Then* $|\{i\} : i \varepsilon N| \leq |B_i : i \varepsilon N|$.

There is, of course, the easy extension:

Theorem 9. *Let* N *be arbitrary and* B_i *finite for all* i, $A_j \cap A_i = \phi$ *for* $j \neq i$, *and* $|A_M| \leq |B_M|$ *for all finite* M. *Then* $|A_i : i \varepsilon N| \leq |B_i : i \varepsilon N|$.

Proof. Put $B_{ix} = B_i$ for $i \varepsilon N$ and $x \varepsilon A_i$. Let M be finite and $A'_i \subset\subset A_i$ for each $i \varepsilon M$. Then, again (9) holds since this merely means that $|A'_M| \leq |B_M|$ which holds by hypothesis. The truth of (9) for finite M and A'_i enables us to apply Theorem 8, and the proof is completed exactly as that of Theorem 7.

Remark. Let $A_j \cap A_i = \phi$ for $j \neq i$. Then the condition

(10) $$|A_M| \leq |B_M| \text{ for all } M \subset\subset N$$

is strictly weaker than the condition

(11) $$|A_i : i \in N| \lesssim |B_i : i \in N|.$$

Consider the following case. $|N| > |B| = \aleph_0$; $A_i = \{i\}$ and $B_i = B$ for all i. Then (10) holds but (11) is false because of $|A_N| = |N| > \aleph_0 = |B_N|$.

We are now going to free ourselves from the restriction that $A_j \cap A_i = \phi$ for $j \neq i$.

Theorem 10. Let either N be finite or B_i be finite for all i. Then

(12) $$|A_i : i \in N| \leq |B_i : i \in N|$$

if and only if

(13) $$|\bigcup(k \varepsilon L) A_{[M_k]}| \leq |\bigcup(k \varepsilon L) B_{[M_k]}|$$

whenever

$$|L| < \aleph_0; \ |M_k| < \aleph_0 \text{ for } k \in L;$$
$$M_k \neq M_{k'} \qquad \text{for } k \neq k'.$$

This result was proved in [2]. The proof that follows is simpler.

Proof. Let (14) hold. Then

$$|\bigcup(k\epsilon L)B_{[M_k]}| \geq |\bigcup(k\epsilon L)A_{[M_k]}| \geq |\bigcup(k\epsilon L)\alpha_{M_k}|$$
$$= \sum(k\epsilon L)|\alpha_{M_k}|.$$

We observe that if N is finite then there are only finitely many choices for the M_k, and if all B_i are finite then every $B_{[M_k]}$ is finite. Hence, by Theorem 7 or Theorem 9, we can find, for each $M \subset N$ a set $h(M) \subset B_{[M]}$ such that $|h(M)| = |\alpha_M|$ for all M, and $h(M) \cap h(M') = \phi$ for $M \neq M'$. Put, for each i, $B_i' = \bigcup(M \ni i)h(M)$. Then $B_i' \subset \bigcup(M \ni i)B_{[M]} \subset B_i$.

Consider a fixed M. By Theorem 3, $h(M) = \beta_M'$, and we have $|\alpha_M| = |h(M)| = |\beta_M'|$. By Theorem 4, $|A_i : i \epsilon N| = |B_i' : i \epsilon N| \leq |B_i : i \epsilon N|$ which completes the proof.

3. Reconstruction theorems. In this section, we shall be concerned with the reconstruction of the isomorphism type of a family from the isomorphism types of its proper subfamilies. Problems in this field may be considered as analogue of the problem posed by Ulam's conjecture on the reconstruction of the isomorphism type of a graph from the isomorphism types of its maximal proper subgraphs. There is, however, a significant difference between the two situations. In Ulam's case, isomorphism from one

graph to another involves a bijection of one node set on another accompanied by a bijection of one edge set on another, both functions being linked by the condition that adjacency and non-adjacency be preserved. In our case, we are concerned with a bijection of one node set, A_N onto the other node set, B_N, which takes every edge A_i onto the edge B_i carrying the same index i. In other words, our problem concerns hypergraphs with labelled edges whereas there are no labels in Ulam's case. We shall prove positive results, in which the analogue of Ulam's conjecture is true, as well as negative results in which it is false. The work contained in this section was carried out in collaboration with C. Berge.

A pair of distinct N-families $(A_i : i \in N)$, $(B_i : i \in N)$ is called a <u>non-reconstructible</u> pair if

(15) $\quad |A_i : i \in N - \{k\}| = |B_i : i \in N - \{k\}|$ for all $k \in N$.

Clearly, (15) implies that

$$|A_i : i \in M| = |B_i : i \in M| \quad \text{for all } M \neq N.$$

We begin by defining, for every finite non-empty N, a pair of nonreconstructible N-families K(N), L(N) whose nodes are non-empty subsets of N. For finite sets X,Y write

$$X \equiv Y$$

whenever

$$|X| \equiv |Y| \pmod 2,$$

and $X \not\equiv Y$ whenever $|X| \not\equiv |Y| \pmod 2$. Consider the N-families

$$K(N) = (K_i(N) : i \in N); L(N) = (L_i(N) : i \in N)$$

where, for each i,

$$K_i(N) = \{M : M \equiv N; i \in M\}; L_i(N) = \{M : M \not\equiv N; i \in M\}.$$

We have, for instance $K(\{1\}) = (\{\{1\}\}); L(\{1\}) = (\{\phi\})$

$$K(\{1,2\}) = (\{\{1,2\}\},\{\{1,2\}\}); L(\{1,2\}) = (\{\{1\}\},\{\{2\}\}),$$

$$K(\{1,2,3\}) = (\{\{1\},\{1,2,3\}\},\{\{2\},\{1,2,3\}\},\{\{3\},\{1,2,3\}\}),$$

$$L(\{1,2,3\}) = (\{\{1,2\},\{1,3\}\},\{\{1,2\},\{2,3\}\},\{\{1,3\},\{2,3\}\}).$$

The families K({1,2,3}), and L({1,2,3}) correspond to the graphs of Figure 1. We often write K and K_i instead of K(N) and $K_i(N)$, and similarly for L.

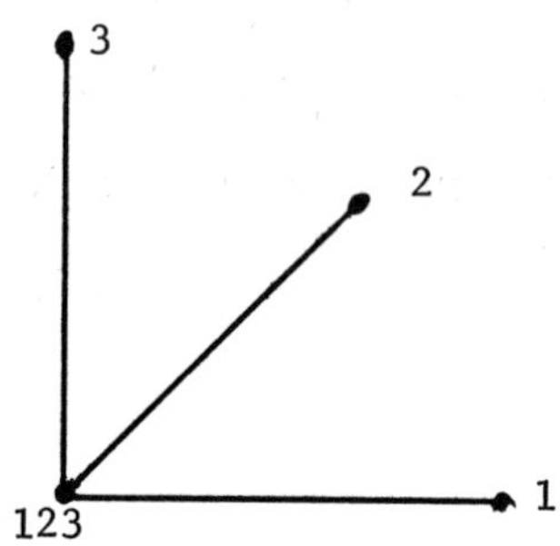

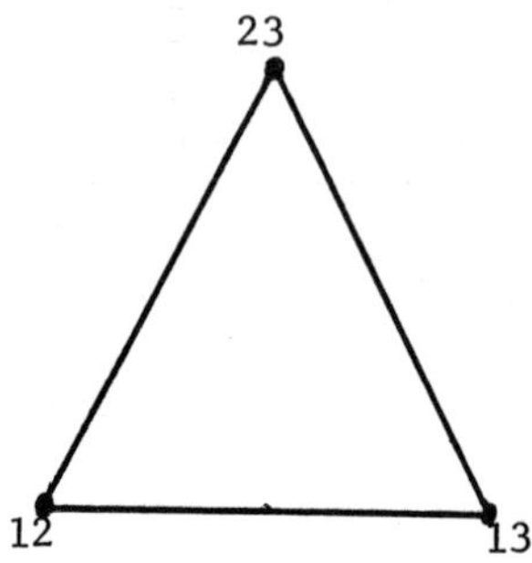

Figure 1

Theorem 11. Let N be finite, $|N| = \eta \geq 1$. Then

(i) $|K_N| = 2^{n-1} - 1$; $|L_N| = 2^{n-1}$ if η is even.

$|K_N| = 2^{n-1}$; $|L_N| = 2^{n-1} - 1$ if η is odd.

(ii) The atoms κ_M, λ_M of K, L are given by

$\kappa_M = \{M\}$, $\lambda_M = \phi$ if $M \equiv N$; $\kappa_M = \phi$; $\lambda_M = \{M\}$ if $M \not\equiv N$.

(iii) For every $k \in N$,

$$|K_i : i \in N - \{k\}| = |L_i : i \in N - \{k\}|.$$

(iv) $|K_i : i \in N| \neq |L_i : i \in N|$.

Proof of (i). This follows from our definitions.

Proof of (ii).

$$\kappa_M = \bigcap (i \in M)\{M' : i \in M' \equiv N\} - \bigcup (i \notin M)\{M' : i \in M' \equiv N\}$$

$$= \{M' : M \subset M' \equiv N\} - \{M' : M \not\supset M' \equiv N\} = \begin{cases} \{M\} & \text{if } M \equiv N \\ \phi & \text{if } M \not\equiv N. \end{cases}$$

For λ_M the same holds except that $\equiv$ and $\not\equiv$ are interchanged.

Proof of (iii). For $X \subset N$ put $f(X)=(X-\{k\}) \cup (\{k\}-X)$. Then $f(f(X)) = X$, and f permutes the subsets of N. Let $i \neq k$ and $M \in K_i$. Then $i \in M \equiv N$ and hence $i \in f(M) \not\equiv N$; $f(M) \in L_i$; $f(K_i) \subset L_i$. By symmetry, $f(L_i) \subset K_i$. Hence $L_i = f(f(L_i)) \subset f(K_i) \subset L_i$, and $f(K_i) = L_i$ for all $i \neq k$.

Proof of (iv). By (ii), $|K_N| \neq |\lambda_N|$. This completes the proof.

We will now give two further definitions of non-reconstructible pairs of N-families applicable to any index set N. The first is an adaptation of the definition of K(N),L(N) which is valid both for finite and for infinite N. When N is finite, it reduces to the construction just established. Since for infinite N it depends on Zorn's lemma, it is non-constructive. Moreover, for infinite N, it does not give a unique pair but one of many pairs. By way of contrast, the second definition is completely explicit, even for infinite N. It lacks, however, the symmetry present in the first definition.

We need a lemma which establishes the notion of parity for subsets of N. For sets X,Y we put

$d(X,Y) = |X - Y| + |Y - X|$.

Lemma. *Put* $\Omega = \{X : X \subset N\}$. *Then there are sets* Ω', Ω'' *such that*

$$\Omega = \Omega' \cup \Omega''; \; \Omega' \cap \Omega'' = \phi.$$

If $X,Y \in \Omega'$ *or* $X,Y \in \Omega''$ *then* $d(X,Y) \notin \{1,3,5,\ldots\}$.

Proof. By Zorn's lemma, there is a maximal set $\Omega' \subset \Omega$ such that $d(X,Y) \notin \{1,3,5,\ldots\} = Q$, say, for all $X,Y \in \Omega'$. Put $\Omega'' = \Omega - \Omega'$. If $X,Y \in \Omega''$ and $d(X,Y) \in Q$ then, by the maximality of Ω', there are sets $X',Y' \in \Omega'$ such that $d(X,X'), d(Y,Y') \in Q$. Then

$$Q \not\ni d(X',Y') \equiv d(X',X) + d(X,Y) + d(Y,Y') \equiv 1 \pmod 2$$

which is a contradiction. This completes the proof.

If N is finite then, by (16), the relation $X \equiv Y$ holds if and only if either $X,Y \in \Omega'$ or $X,Y \in \Omega''$. This justifies the introduction of the following definition. Let N be infinite. We make a choice of a set Ω' as mentioned in the lemma, and for $X,Y \subset N$ we put $X \equiv Y$ if either $X,Y \in \Omega'$ or $X,Y \in \Omega''$. We now put, for every N,

$$K = (K_i : i \in N); \; L = (L_i : i \in N)$$

where, for each i, $K_i = \{M : i \in M \equiv N\}$; $L_i = \{M : i \in M \not\equiv N\}$.

Theorem 11'. The statements (ii),(iii),(iv) of Theorem 11 hold for every set N. If N is infinite, then for every i, $|K_i| = |L_i| = 2^{|N|}$.

Proof. Let N be infinite. The proofs of (ii)-(iv) are identical with the proofs of the same propositions for finite N, the essential point being that, by (16), $X \not\supseteq f(X)$ for every X, where f is the function used in the proof of (iii) for finite N. There remains the proof of the final clause. Let $i \in N$. Then we can choose $k \in N - \{i\}$. Put, for $X \subset N$, $h(X) = (X - \{k\}) \cup (\{k\} - X)$. Then $h(h(X)) = X$, and h is injective. By (16), if $i \in X$ then $i \in h(X) \not\supseteq X$. Hence $h(K_i) \subset L_i; h(L_i) \subset K_i$ and, if N is infinite, $|K_i| = |L_i| = |K_i \cup L_i| = 2^{|N|}$.

Let us now turn to the explicit, though non-symmetrical construction.

Theorem 11". Let $|N| \geq 2$ and $i_1 \in N$. Let, for $i \in N$ D_i be a set such that $0 < |D_{i_1}| \leq |D_i| \geq \aleph_0$ for $i \in N - \{i_1\}$, and $D_k \cap D_i = \phi$ for $k \neq i$. Put $A_{i_1} = D_N$; $B_{i_1} = D_{N-\{i_1\}}$,

$A_i = B_i = D_i$ for $i \in N - \{i_1\}$. Then

(i) $|A_i : i\varepsilon N-\{i_o\}| = |B_i : i\varepsilon N-\{i_o\}$ for $i_o \in N$,

(ii) $|A_i : i\varepsilon N| \neq |B_i : i \in N|$.

Proof of (i). Case 1. $i_o \neq i_1$. Since D_{i_o} is infinite and $|D_{i_1}| \leq |D_{i_o}|$ there is a bijection $f : D_{i_o} \cup D_{i_1} \to D_{i_o}$. Define a function g on D_N by putting $g(x) = f(x)$ for $x \in D_{i_o} \cup D_{i_1}$; $g(x) = x$ for $x \in D_{N-\{i_o,i_1\}}$. Clearly, g is injective. We have

$$g(A_{i_1}) = g(D_N) = g(D_{i_o} \cup D_{i_1}) \cup g(D_{N-\{i_o,i_1\}})$$

$$= D_{i_o} \cup D_{N-\{i_o,i_1\}} = D_{N-\{i_1\}} = B_{i_1}.$$

Also, if $i \in N - \{i_o,i_1\}$ then $g(A_i) = g(D_i) = B_i$.

Case 2. $i_o = i_1$. Then put $g(x) = x$ for $x \in D_{N-\{i_1\}}$. Then $g(A_i) = B_i$ for $i \neq i_o$, and (i) follows.

Proof of (ii).

$$|A_{i_1} - A_{N-\{i_1\}}| = |D_{i_1}| > 0 = |B_{i_1} - B_{N-\{i_1\}}|.$$

It might be of interest to exhibit the pairs α,β of corresponding atoms of the two families occurring in (i). For $M \subset N - \{i_o\}$ put

$$\alpha = A_{[M]} - A_{(N-\{i_o\})-M};\ \beta = B_{[M]} - B_{(N-\{i_o\})-M};$$

$$t = |M \cap (N-\{i_1\})|;\ \gamma_o = D_{[M]} - D_{(N-\{i_1\})-M};$$

$$\gamma_1 = D_{i_o} \cup D_{i_1};\ \gamma_2 = D_{i_o}.$$

We had already noted that $|\gamma_1| = |\gamma_2|$. In the diagram of Figure 2, the values of α, β are written against the end nodes corresponding to the various cases.

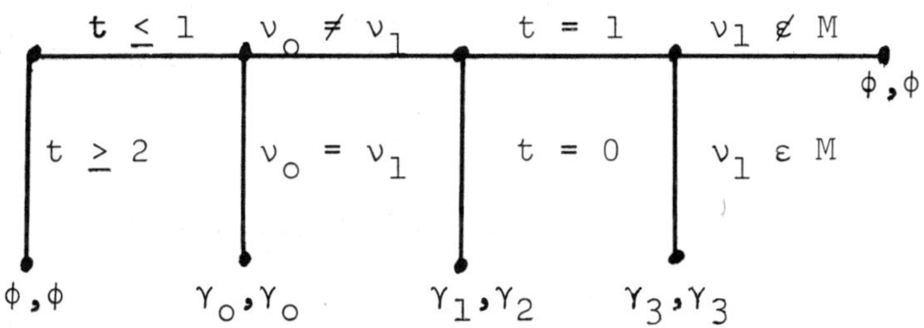

Figure 2

We shall now consider positive theorems which assert reconstructibility of a family from its subfamilies. In our first theorem of this kind the isomorphism types of all finite subfamilies already enable us to reconstruct the type of the whole family, provided the members of the family are finite. One might suspect that such a theorem could be established with the help of some general compactness principle such as Tychonoff's Theorem, but it appears that such arguments only yield considerably weaker results than the method used below for proving Theorem 12. Let us remember that $X \subset\subset Y$ means that X is finite and $X \subset Y$.

Theorem 12. *Suppose that, for every* $M \subset\subset N$,

(18) $$(A_i : i \in M) \cong (B_i : i \in M),$$

and let

(19) $$|A_i| = |B_i| < \aleph_0 \quad \textit{for all } i \in N.$$

Then

(20) $$(A_i : i \in N) \cong (B_i : i \in N).$$

Remark 1. The proof will show that the hypothesis (19) may be replaced by the following weaker condition:

(19') *for every non-empty subset* M_o *of* N *the set* $A^*(M_o) = \{A_{[R]} - A_S : M \subset\subset M_o;\ S \subset\subset N - M_o\}$ *contains an inclusion-minimal member and similarly with* B *in place of* A.

Clearly (19) implies that all members of $A^*(M_o)$ and $B^*(M_o)$ are finite so that (19') is true in this case.

Remark 2. The hypothesis (19) cannot be omitted altogether as is shown by the following case.

$$N = \{1,2,\ldots\};\ A_1 = \{a_1,a_2,\ldots\};\ B_1 = \{b_1,b_2,\ldots\};$$

$A_i = \{a_{i-1}\}$ and $B_i = \{b_i\}$ for $i \geq 2$;

$a_\mu \neq a_i$ and $b_\mu \neq b_i$ for $\mu \neq i$.

We will now show that (18) holds if and only if either (i) $1 \notin M$ or (ii) $1 \in M$ and $|N - M| = \aleph_0$. In particular, (18) is true but (20) is false.

If $1 \notin M$ then the map $a_i \to b_{i+1} (i \geq 1)$ proves (18). Now let $1 \in M$ and $|N - M| = \aleph_0$. Define $f : A_M \to B_M$ by putting, first of all, $f(a_{i-1}) = b_i$ for $i \in M - \{1\}$. Then $f(A_i) = B_i$ for $i \in M - \{1\}$. Since each of the sets $A_1 - \{a_{i-1} : i \in M - \{1\}\}$, $B_1 - \{b_i : i \in M - \{1\}\}$ is denumerable, we can complete the definition of f in such a way that f is bijective and $f(A_1) = B_1$. This shows that (18) holds if (i) or (ii) is satisfied. Now assume that neither (i) nor (ii) holds. Then $1 \in M$ and $|N - M| < \aleph_0$. Suppose that there is a bijection $g : A_M \to B_M$ such that $g(A_i) = B_i$ for all $i \in M$. Then $g(a_{i-1}) = b_i$ for $i \in M - \{1\}$. Hence there is $k > 2$ such that $g(a_{i-1}) = b_i$ for every $i \geq k$. But then g provides a one-to-one map of $\{a_1, a_2, \ldots, a_{k-2}\}$ onto $\{b_1, b_2, \ldots, b_{k-1}\}$ which is impossible.

<u>Proof of Theorem 12</u>. Let us put, for $M \subset N$ and $S \subset N$,

$$A(M,S) = A_{[M]} - A_S; \quad B(M,S) = B_{[M]} - B_S.$$

We observe that if $M \subset M'$ and $S \subset S'$, then

$$A(M,S) \supset A(M',S'),$$

and similarly for B. Now let $M_o \subset N$. By (19'), we can find an inclusion-minimal member $A(M_1,S_1)$ of $A^*(M_o)$, where $M_1 \subset\subset M_0$ and $S_1 \subset\subset N - M_o$. We have

$$A(M_1,S_1) \supset A(M_o,N - M_o).$$

Let us assume that there is an element

$$x_o \in A(M_1,S_1) - A(M_o,N - M_o).$$

If $x_o \notin A_{[M_o]}$ then $x_o \notin A_{i_1}$ for some $i_1 \in M_o$. Then by the minimality, $x_o \notin A(M_1 \cup \{i_1\},S_1) = A(M_1,S_1)$ which is a contradiction. Hence $x_o \in A_{[M_o]}$. Then $x_o \in A_{N-M_o}$, and we have $x_o \in A_{i_2}$ for some $i_2 \in N-M_o$. Again, by minimality,

$$x_o \notin A(M_1,S_1 \cup \{i_2\}) = A(M_1,S_1)$$

which is a contradiction. Thus there is no such element x_o, and we have

$$A(M_1,S_1) = A(M_o,N - M_o).$$

Now, using (18) and $|M_1 + S_1| < \aleph_0$, we find

$$|A(M_0,N-M_0)|=|A(M_1,S_1)|=|B(M_1,S_1)|\geq|B(M_0,N-M_0)|.$$

By symmetry, $|B(M_0,N - M_0)| \geq |A(M_0,N - M_0)|$.

Corresponding atoms of the two families in (18) have the same cardinal, and the theorem follows.

In conclusion, I state two theorems which show that if N and the A_i and B_i are finite, the special non-reconstructible pair K(N),L(N) is always involved in a certain way whenever the phenomenon of non-reconstructibility occurs.

Theorem 13. *Let* N *be finite*, $|N| = n \geq 1$. *Let* A_i, B_i, X, Y *satisfy the following conditions:* $|X| = |Y| < \aleph_0$; $A_N \subset X$; $B_N \subset Y$; $(A_i : i \in N - \{k\}) \cong (B_i : i \in N - \{k\})$ *for all* $k \in N$; *there are no sets* A,B *with* $A \subset X$; $B \subset Y$; $|A| = |B| = 2^{n-1}$ *such that the families* $(A_i \cap A : i \in N)$, $(B_i \cap B : i \in N)$, *in any order, are isomorphic to* K(N),L(N) *respectively. Then*

$$(A_i : i \in N) \cong (B_i : i \in N).$$

We should like to remark that if there is a set A satisfying

$$(21) \qquad (A_i \cap A : i \in N) \cong K$$

then there is also a set A' such that $A' \subset A_N$; $|A'| = |K_N(N)|$ and

$$(A_i \cap A' : i \in N) \cong K.$$

Indeed, (21) implies

$$|A_N \cap A| = |\bigcup(i \in N) A_i \cap A| = |\bigcup(i \in N) K_i| = |K_N(N)|,$$

so that the set $A' = A_N \cap A$ has the desired property. A similar remark applies so that the set $A' = A_N \quad A$ has the desired property. A similar remark applies to L.

The proof of Theorem 13 will appear in a forthcoming paper [6]. All we shall do with Theorem 13 here is to show that there is A satisfying (21) if and only if

$$(22) \qquad \alpha_M \neq \phi \qquad \text{for all } M \equiv N.$$

Let (21) hold. Put $A_i \cap A = A_i'$. Then $M \equiv N$ implies

$$|\alpha_M \cap A| = |\alpha_M'| = |\kappa_M| \neq 0$$

and therefore (22). Vice versa, if (22) is true then we can choose $x_M \in \alpha_M$ for all M such that

$M \equiv N$. Put $A = \{x_M : M \equiv N\}$. Then $A_i \cap A = \{x_M : i \in M \equiv N\}$, and (21) holds.

The following theorem is more powerful than Theorem 13 in that it incorporates a whole sequence of similar propositions, one for each value of the parameter p.

Theorem 14. Let p *be a positive integer and* $|N| = n \geq p$. *Let* A_i, B_i *be finite and* $(A_i : i \in N') \cong (B_i : i \in N')$ *for every set* $N' \subset N$ *such that* $|N'| = p-1$. *Suppose that there are no sets* A, B, P *with* $A \subset A_N$; $B \subset B_N$; $P \subset N$ *such that* $|P| = p$, *and the two families* $(A_i \cap A : i \in P)$, $(B_i \cap B : i \in P)$, *in any order, are isomorphic to* $K(P), L(P)$ *respectively. Then*

$$(A_i : i \in N) \cong (B_i : i \in N). \tag{23}$$

We shall now deduce Theorem 14 from Theorem 13. Let p be fixed. Put $X = Y = A_N \cup B_N$. First of all, let $n = p$. We observe that we have, for every $A \subset X$, the relations $A_i \cap A = A_i \cap (A_N \cap A)$ and $A_N \cap A \subset A_N$. Hence Theorem 13 applies and yields (23). Next, let $n > p$, and use induction with respect to n. Then, by the induction hypothesis

$$(A_i : i \in N - \{k\}) = (B_i : i \in N - \{k\}) \tag{24}$$

for every $k \in N$. We now want to show that there are no sets A^*, B^* such that $A^* \subset X$ and $B^* \subset Y$, and the families

$$(A_i \cap A^* : i \in N), \ (B_i \cap B^* : i \in N),$$

in any order, are isomorphic to $K(N), L(N)$ respectively. Suppose there are such A^*, B^* and that, for instance,

$$(A_i \cap A^* : i \in N) \cong K(N); \ (B_i \cap B^* : i \in N) \cong L(N).$$

Then there is a bijection $f : A_N \cap A^* \to K_N(N)$ such that $f(A_i \cap A^*) = K_i(N)$ for all $i \in N$. Choose $P \subset N$ such that $|P| = p$. Then $f(A_p \cap A^*) = \bigcup(i \in P) f(A_i \cap A^*) = \bigcup(i \in P) K_i(N) = K_P(N)$

Case 1. $P \equiv N$. Then $K_P(P) \subset K_P(N)$, and there is a set $A \subset A_P \cap A^*$ such that $f(A) = K_P(P)$. Then, for every $i \in P$,

$$f(A_i \cap A) = f((A_i \cap A^*) \cap A) = f(A_i \cap A^*) \cap f(A)$$
$$= K_i(N) \cap K_P(P) = K_i(P).$$

This shows that $(A_i \cap A : i \in P) \cong K(P)$. A similar argument yields a set $B \subset B_P \cap B^*$ such that $(B_i \cap B : i \in P) \cong L(P)$. This violates the hypothesis of the theorem.

Case 2. $P \not\equiv N$. Then $L_P(P) \subset K_P(N)$, and there is a set $A \subset A_P \cap A^*$ such that $f(A) = L_P(P)$. Then, for $i \in P$, $f(A_i \quad A) = f((A_i \quad A^*) \quad A)$
$= f(A_i \cap A^*) \cap f(A) = K_i(N) \cap L_P(P) = L_i(P)$. Hence $(A_i \cap A : i \in P) \cong L(P)$. Similarly we can find a set $B \subset B_P \cap B^*$ such that $(B_i \cap B : i \in P) \cong K(P)$. This, again, violates the hypothesis. Hence there are no sets A^*, B^* with the properties described above, and in view of (24) it follows from Theorem 13 that (23) is true. This completes the deduction of Theorem 14 from Theorem 13.

The cases $p \leq 3$ of Theorem 14 imply some theorems of Whitney [4] and Sabidussi [5].

REFERENCES

[1] P. Hall, On representatives of subsets, J. London Math. Soc. 10 (1935), 26-30.

[2] R. Rado, A theorem on general measure functions, Proc. London Math. Soc. 44 (1938), 61-91.

[3] M. Hall Jr., Distinct representatives of subsets, Bull. Amer. Math. Soc. 54 (1948), 922-928.

[4] H. Whitney, Congruent graphs and the connectivity of graphs, Amer. J. Math. 54 (1932), 150-168.

[5] G. Sabidussi, Graph derivatives, Math. Zeitschr. 76 (1961), 385-401.

[6] C. Berge and R. Rado, Note on isomorphic hypergraphs and some extensions of Whitney's theorem to families of sets, to appear.

[7] C. Berge, Graphes et Hypergraphes, Dunod, Paris, 1970.

COUNTING LABELED ACYCLIC DIGRAPHS

Robert W. Robinson

1. Introduction.

A *digraph* is a finite directed graph with no loops or multiple arcs. In this paper, all digraphs considered have labeled points, but arcs and other subgraphs are never labeled. Lower case Greek letters will denote digraphs. Upper case Latin letters will denote sets of digraphs. Isomorphic digraphs are not to be distinguished, so for instance a sum over all δ in S is to include just one representative of each isomorphism class of members of S.

The number of points of any digraph δ is denoted by $p(\delta)$. A *labeling* of δ is a linear ordering of its points. A linear ordering corresponds to the assignment of the numbers $1,2,\ldots,p(\delta)$ to the points of δ for which the label i is assigned to the i'th point in the ordering. Any digraph γ is isomorphic to δ only if $p(\gamma) = p(\delta)$. In that case, there is a unique 1-1 correspondence of the points of δ with those of γ which preserves the labeling, namely, the two points labeled i must correspond to each other for all $1 \leq i \leq p(\delta)$. Then γ and δ are isomorphic just if this correspondence preserves the adjacency relation, i.e., the arcs. Any subgraph α of δ is labeled by the restriction of the linear ordering on the points of δ. Thus there may be no agreement between the corresponding assignment of $1,2,\ldots,p(\alpha)$ for α with the assignment of $1,2,\ldots,p(\delta)$ for δ.

Standard terminology for digraphs and connectedness is established in the book [7] of Harary, Norman and Cartwright. For the sake of completeness, recall that a digraph is _weakly connected_ (or _weak_) if its underlying graph is connected, and is _strongly connected_ (or just _strong_) if every point can be reached from every other (by a directed path). An exception to this definition is the empty digraph, which is neither weak nor strong. Note that the empty digraph is also called the null graph [10].

Any digraph can be divided into weak or strong components, which are the maximal weakly or strongly connected subgraphs. The strong components of a digraph partition the points but do not in general include all of the arcs. The weak components of a digraph partition both the points and the arcs. For this reason, it is straightforward to enumerate digraphs with given weak components. The result is presented in terms of exponential generating functions later in the section.

The more delicate matter of enumerating digraphs with given strong components requires a different sort of generating function, and is the subject of Section 2. In Section 3, the results of Section 2 are used to find the number of acyclic digraphs, and to obtain strict asymptotic estimates of these numbers. Results of Gilbert [4] and Wright [18] are applied to obtain similar results for weak acyclic digraphs.

An _out-component_ of a digraph is a strong component which cannot be reached from any other strong component. In case an out-component is a single point, the terminology _out-point_ may be used. The next lemma is basic to the main theorem.

Lemma 1. Every non-empty digraph has an out-component.

The proof is omitted because of its

simplicity; see [7, pp. 57-65].

Obviously a digraph is acyclic just if every strong component is a single point. This makes it easy to apply the main theorem to count acyclic digraphs. Our earlier method of enumerating acyclic digraphs [16] suffers from the necessity of maintaining the number of out-points as an enumeration parameter as well as the total number of points. This sort of information can also be obtained from the results of Section 2, where theorems similar to the main one count digraphs with given strong components and any fixed number of out-components. In the case of acyclic digraphs, strict asymptotic results are obtained for the number which have n out-points, as a function of the number p of points. It turns out that for each $n \geq 1$, the fraction of the acyclic digraphs on p points which have n out-points approaches a positive constant as p grows large.

In Section 4, the theorems of Section 2 are applied to count strong digraphs and digraphs with a source. This leads to a natural derivation of the relations of Wright [19] for strong digraphs. Liskovec [12] earlier counted digraphs with a rooted source as part of a more complicated set of relations for strong digraphs, on which the work of Wright is based. In Section 4, it is also indicated how to count unilaterally connected digraphs. The technique suffers from the defect that the number of points in the (unique) out-component must be an enumeration parameter, as well as the total number of points. Section 5 comments on related problems, comparing those which have been solved with those which remain open.

We now introduce the types of generating functions which are to be employed, and give a result which will be used several times later.

For any set S of digraphs , the <u>exponential generating function</u> S(x) is defined by

$$S(x) = \sum_{\delta \varepsilon S} \frac{x^{p(\delta)}}{p(\delta)!} \quad . \tag{1}$$

Note that the number s_i of digraphs in S with i points is just i! times the coefficient of x^i in S(x).

The next notation is essential. Let Δ be the linear operation on generating functions which divides x^i by $2^{\binom{i}{2}}$. Thus

(2) $$\Delta S(x) = \sum_{\delta \varepsilon S} \frac{x^{p(\delta)}}{p(\delta)!\, 2^{\binom{p(\delta)}{2}}} .$$

We call $\Delta S(x)$ the <u>special generating function</u> for S. We denote by Δ^{-1} the operation which is inverse to Δ. The use of exponential or special generating functions gives multiplication of the functions properties useful to different applications. The choice of generating function for various purposes has been studied in some generality by Bender and Goldman [1]. In particular, exponential generating functions are developed in their Example 5, and a type equivalent to special generating functions is developed in their Example 6. Sometimes it is desirable to enumerate digraphs with respect to other parameters than the number of points. For any digraph δ, let $q(\delta)$ be the number of arcs of δ and $c(\delta)$ the number of weak components of δ. We now define the two <u>extended generating functions</u> for S (with respect to arcs and weak components) by

(3) $$S(x,y) = \sum_{\delta \varepsilon S} \frac{x^{p(\delta)} y^{q(\delta)}}{p(\delta)!}$$

and

$$(4) \qquad S(x,z) = \sum_{\delta \varepsilon S} \frac{x^{p(\delta)} z^{c(\delta)}}{p(\delta)!} .$$

Thus x is the point-variable, y the arc-variable, and z the weak-component variable. Since the points of δ are labeled but nothing else is, an exponential factor $p(\delta)!$ is introduced just for the points. The special operator Δ likewise is defined to act only on powers of x. Thus for instance

$$(5) \qquad \Delta S(x,y) = \sum_{\delta \varepsilon S} \frac{x^{p(\delta)} y^{q(\delta)}}{p(\delta)!\, 2^{\binom{p(\delta)}{2}}}$$

Lemma 2. If S is a set of weakly connected digraphs and W_S is the set of all digraphs whose weak components lie in S, then

$$W_S(x,z) = e^{S(x,z)}$$

The lemma is also true with powers of y for the arc counting parameter in addition to the others. The idea of the lemma has appeared in the literature often enough to give it the status of folklore. The first explicit statement sufficiently general to justify the lemma seems to be due to Uhlenbeck and Ford [17, Theorem I, p. 40]. The necessary methods can be found also in the book of

Harary and Palmer [9, Chapter 1], or in Bender and Goldman [1, Example 5].

As in the lemma, suppose S is a set of weak digraphs. The notation W_S for the set of all digraphs with weak components in S will be adopted for the remainder of the paper. Letting $W_S^{(n)}$ be the set of digraphs in W_S which have exactly n weak components, we have

$$W_S(x,z) = \sum_{n=0}^{\infty} z^n W_S^{(n)}(x). \tag{6}$$

On the other hand,

$$S(x,z) = zS(x)$$

since S consists of weak digraphs. Equating coefficients of z^n from (6) and from Lemma 2 gives the useful fact

$$W_S^{(n)}(x) = \frac{1}{n!} S(x)^n, \tag{7}$$

which does not appear to lie in the literature even though it has an easy direct proof.

2. Digraphs with given strong components.

In this section, C is a set of labeled strong digraphs and D_C is the set of all labeled digraphs such that every strong component is a member of C.

Theorem 1. $\Delta D_C(x) = (\Delta e^{-C(x)})^{-1}$.

Proof. As above let W_C be the set of all labeled digraphs with every weak component in C. For convenience, we define an out-rooted digraph as one with a distinguished set of out-components, possibly empty; these are called the root components. Let R_C be the set of out-rooted digraphs such that every strong component is in C. For any out-rooted δ let $r(\delta)$ be the number of root components, $c(\delta)$ the number of strong components, and $p(\delta)$ the number of points in δ.

We first show that

$$(8) \qquad \sum_{\delta \varepsilon R_C} (-1)^{r(\delta)} \frac{x^{p(\delta)}}{p(\delta)!\, 2^{\binom{p(\delta)}{2}}} = 1.$$

To verify the identity (8), fix $\gamma \varepsilon D_C$ and suppose that γ has exactly n out-components. There are $\binom{n}{i}$ ways to choose i of these to be the root components of an out-rooted digraph with γ as the underlying digraph. Thus all of the out-rooted digraphs with underlying digraph γ contribute

$$\sum_{i=0}^{n} (-1)^i \binom{n}{i} \frac{x^{p(\delta)}}{p(\delta)!\, 2^{\binom{p(\delta)}{2}}}$$

to the left hand side of (8). Into this expression, we substitute the well-known identity,

$$\sum_{i=0}^{n} (-1)^i \binom{n}{i} = \begin{cases} 0 \text{ if } n > 0 \\ 1 \text{ if } n = 0. \end{cases}$$

By Lemma 1, the unique digraph in D_C with $n = 0$ out-components is the empty digraph, which has $p = 0$ points, proving (8).

We now apply (8) to prove the theorem by factoring the left side as follows. For every α in R_C, the subgraph induced by the root components is some digraph δ in W_C. For the root components of α are all out-components, and no out-component can be reached from another. Let γ be the subgraph of α induced by the strong components which are not root components. Clearly γ is in D_C. The ordered pair of digraphs (δ,γ) is uniquely determined by α. Given any (δ,γ) in $W_C \times D_C$, so that δ has its weak components in C and γ has its strong components in C, it is easy to see that there are just

$$2^{p(\delta)p(\gamma)} \binom{p(\delta)+p(\gamma)}{p(\delta)}$$

digraphs to which (δ,γ) corresponds in the indicated fashion. For the labelings of the points of δ and

γ may be interlaced in $\binom{p(\delta)+p(\gamma)}{p(\delta)}$ ways, and any arcs joining δ and γ must be directed from δ to γ, giving $p(\delta)p(\gamma)$ independent possibilities for including or excluding an arc. If α corresponds to (δ,γ) then $r(\alpha) = c(\delta)$ and $p(\alpha) = p(\delta) + p(\gamma)$. Thus the left side of (8) can be written

$$\sum_{\delta\varepsilon W_C} \sum_{\gamma\varepsilon D_C} \frac{(-1)^{c(\delta)} x^{p(\delta)+p(\gamma)} 2^{p(\delta)p(\gamma)}}{(p(\delta)+p(\gamma))!\ 2^{\binom{p(\delta)+p(\gamma)}{2}}} \binom{p(\delta)+p(\gamma)}{p(\delta)} .$$

With the two observations that

$$\frac{1}{(p(\delta)+p(\gamma))!} \binom{p(\delta)+p(\gamma)}{p(\delta)} = \frac{1}{p(\delta)!p(\gamma)!} ,$$

$$\binom{p(\delta)+p(\gamma)}{2} - p(\delta)p(\gamma) = \binom{p(\delta)}{2} + \binom{p(\gamma)}{2},$$

the identity (8) becomes

$$(9) \qquad \sum_{\delta\varepsilon W_C} \frac{(-1)^{c(\delta)} x^{p(\delta)}}{p(\delta)!\ 2^{\binom{p(\delta)}{2}}} \sum_{\gamma\varepsilon D_C} \frac{x^{p(\gamma)}}{p(\gamma)!\ 2^{\binom{p(\gamma)}{2}}} = 1.$$

The right factor in (9) is just $\Delta D_C(x)$. If $W_C(x,z)$ is the extended generating function for W_C with respect to points and weak components, then the left factor in (9) can be written $\Delta W_C(x,-1)$. By Lemma 2, we have

$$W_C(x,-1) = e^{-C(x)},$$

so (9) becomes

$$\Delta e^{-C(x)} \Delta D_C(x) = 1,$$

which proves the theorem.

The next result can be proved in a similar way.

Theorem 2. If C is a set of labeled strong digraphs, n a non-negative integer, and $D_C^{(n)}$ the set of labeled digraphs with strong components in C and exactly n out-components, then

$$\Delta D_C^{(n)}(x) = (\Delta e^{-C(x)})^{-1} \Delta\left(\frac{C(x)^n}{n!} e^{-C(x)}\right).$$

We give an outline of how the proof parallels that of Theorem 1. A root 2-colored digraph will be an out-rooted digraph in which each root component is one of two colors, say red or green. Let $R_C^{(n)}$ be the set of root-2-colored digraphs with strong components in C and exactly n root components which are colored red. For every δ in $R_C^{(n)}$ let $s(\delta)$ be the number of root components which are colored green. Then we have

$$\sum_{\delta \varepsilon R_C^{(n)}} \frac{(-1)^{s(\delta)} x^{p(\delta)}}{p(\delta)!\; 2^{\binom{p(\delta)}{2}}} = \Delta D_C^{(n)}(x),$$

by the same sort of argument as for (8). With the

help of (7), the left side can be factored to give

$$\Delta\left(\frac{C(x)^n}{n!} e^{-C(x)}\right) \Delta D_C(x) = \Delta D_C^{(n)}(x).$$

Note that Theorem 1 is obtained as the special case when $n = 0$.

Theorems 1 and 2 are easily extended to include the number of arcs as an enumeration parameter. It is simply a matter of replacing 2 by 1+y whenever the factor of 2 represents the choice of including a possible arc or not. Symbolically let Δ_y be the linear operation on generating functions such that

$$\Delta_y(x^m y^n) = \frac{x^m y^n}{(1+y)^{\binom{m}{2}}}$$

for all m and n. Then we have

$$(10)\quad \Delta_y D_C(x,y) = (\Delta_y e^{-C(x,y)})^{-1},$$

$$(11)\quad \Delta_y D_C^{(n)}(x,y) = (\Delta_y e^{-C(x,y)})^{-1}\ \Delta_y\left(\frac{C(x,y)^n}{n!} e^{-C(x,y)}\right)$$

The strict parallel between Theorems 1 and 2 and equations (10) and (11) makes it unnecessary to point out explicitly for each application how to include the number of arcs as an enumeration parameter when desired.

3. Acyclic digraphs.

We have already observed that a digraph is acyclic just if every strong component is a single point. Let A be the set of acyclic digraphs. Then the exponential counting series for the strong components is just x. We can now specialize Theorem 1 to acyclic digraphs.

Corollary 1. $\Delta A(x) = (\Delta e^{-x})^{-1}$. [This equation was discovered independently by Richard P. Stanley, by way of the interesting fact that if χ_g is the chromatic polynomial of a labeled symmetric graph g on p points, then $(-1)^p \chi_g(-1)$ is the number of acyclic orientations of g.]

Let $f(x) = \Delta e^{-x}$, or more explicitly

$$f(x) = \sum_{i=0}^{\infty} \frac{(-1)^i x^i}{i!\ 2^{\binom{i}{2}}} . \tag{12}$$

Let a_k be the number of acyclic digraphs on k points. The relation $f(x)\Delta A(x) = 1$ gives the recurrence formula

$$a_k = \sum_{i=1}^{k} (-1)^{i+1} \binom{k}{i} 2^{i(k-i)} a_{k-i} , \tag{13}$$

for all $k > 0$, with $a_0 = 1$.

Let $A^{(n)}$ be the set of acyclic digraphs with exactly n out-points, and let $a_k^{(n)}$ be the number with k points. Theorem 2 can now be specialized to cyclic digraphs.

Corollary 2. $\Delta A^{(n)}(x) = (\Delta e^{-x})^{-1} \Delta (\frac{x^n}{n!} e^{-x})$.

From Corollaries 1 and 2, it is straightforward to deduce that

$$(14)\quad a_k^{(n)} = (-1)^{k-n} \binom{k}{n} \sum_{i=0}^{k-n} (-1)^i \binom{k-n}{i} 2^{i(k-i)} a_i$$

holds for all n and k. Of course $a_k^{(n)} = 0$ if $k < n$. The results of applying (13) and (14) for all $k \leq 10$ are given in Table 1.

[The author is grateful to Arnon Rosenthal for devising and running a computer program based on (13) which gave the values of a_n presented in Table 1, and for providing other numerical results which are noted later.]

In Figure 1 are pictured the six different unlabeled acyclic digraphs on three points. Above each is the number of different labelings, and below the number of out-points.

Table 1

The number of acyclic digraphs on ≤ 10 points with given numbers of out-points.

n	a_n	$a_n^{(0)}$	$a_n^{(1)}$	$a_n^{(2)}$	$a_n^{(3)}$	$a_n^{(4)}$	$a_n^{(5)}$
0	1	1	0	0	0	0	0
1	1	0	1	0	0	0	0
2	3	0	2	1	0	0	0
3	25	0	15	9	1	0	0
4	543	0	316	198	28	1	0
5	29,281	0	16,885	10,710	1,610	75	1
6	3,781,503	0	2,174,586	1,384,335	211,820	10,575	186
7	1,138,779,265	0	654,313,415	416,990,763	64,144,675	3,268,125	61,845
8	783,702,329,343	0	450,179,768,312	286,992 935,964	44,218 682,312	2,266,772,550	43,832,264
9	1,213,442,454 842,881	0	696,979,588 034,313	444,374,705 175,516	68,501,035 223,124	3,518,110 987,650	68,475 488,382
10	4,175,098,976 430,598,143	0	2,398,044,825 254,021,110	1,528,973,599 758,889,005	235,728,863 806,525,320	12,113,567 232,829,650	236,244,835 620,612

Table 1 (continued)

The number of acyclic digraphs on ≤ 10 points with given numbers of out-points.

n	$a_n^{(6)}$	$a_n^{(7)}$	$a_n^{(8)}$	$a_n^{(9)}$	$a_n^{(10)}$
0	0	0	0	0	0
1	0	0	0	0	0
2	0	0	0	0	0
3	0	0	0	0	0
4	0	0	0	0	0
5	0	0	0	0	0
6	1	0	0	0	0
7	441	1	0	0	0
8	336,924	1016	1	0	0
9	538,180,524	1,751,076	2,295	1	0
10	1,869,313 044,690	6,220,861,320	8,801,325	5,110	1

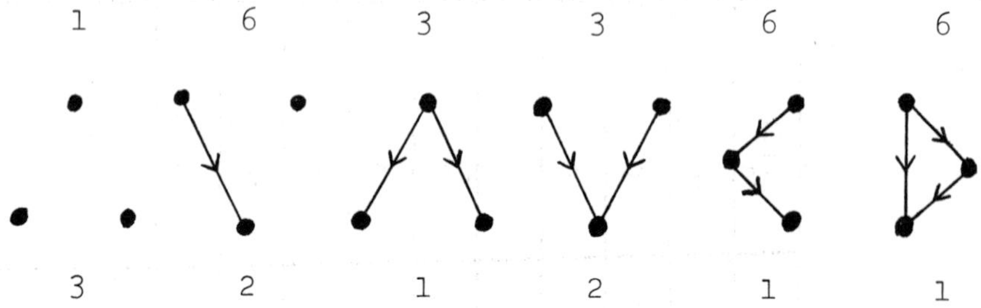

Figure 1. The acyclic digraphs on three points.

Asymptotic estimates for a_k and $a_k^{(n)}$ are obtained by considering the zeros of f(x), where (12) is viewed as a power series definition of f as a function of the complex variable x. [Thanks are due to Peter Rosmarin and Keith Miller for useful discussions on complex variables, which served to set the author on the right track.] It follows from (12) by a result of Hadamard [5] that f is an entire function of order 0. Thus if $z_0, z_1, z_2, \ldots$ denote the zeros of f, say with increasing modulus, then f can be written as

$$f(x) = \prod_{i=0}^{\infty} \left(1 - \frac{x}{z_i}\right). \tag{15}$$

We now need to show that the zeros of f are all real. This is accomplished by invoking classical results. [The author is grateful to N. G. de Bruijn for the references given here.] In [14], Pólya and Schur called a sequence

$q_0, q_1, q_2, \ldots$ a <u>factor sequence of the first kind</u> if, whenever the polynomial

$$b_0 + b_1 x + \ldots + b_m x^m$$

has only real zeros, the same is true of

$$b_0 q_0 + b_1 q_1 x + \ldots + b_m q_m x^m.$$

They showed [14, p. 104] that if $q_0, q_1, q_2, \ldots$ is a factor sequence of the first kind, then the power series

$$q_0 + q_1 x + \ldots + \frac{q_m x^m}{m!} + \ldots$$

defines an entire function whose zeros are all real and of the same sign. Laguerre [11, p. 133] had already shown that if $0 < t < 1$, then

$$1, t, t^3, \ldots, t^{\binom{m+1}{2}}, \ldots$$

is a factor sequence of the first kind. It is easy to see that

$$1, 1, t, t^3, \ldots, t^{\binom{m}{2}}, \ldots$$

must also be a factor sequence of the first kind, so the theorem of Pólya and Schur applies for $t = 1/2$ to show that the zeros $z_0, z_1, z_2, \ldots$ of f are all real.

It is evident that, being real, each z_i is positive. From (12), it is not hard to check that f satisfies the functional equation

$$f'(x) = -f(x/2). \tag{16}$$

It is curious that the functional equation satisfied by f(-x), which is the same as (16) without the minus sign, was studied by Mahler [13] in connection with the problem of enumerating partitions of integers into powers of 2. From (16), it is seen that f attains its first relative extremum after z_i at $2z_i$, so the next zero is $>2z_i$. Consequently $f'(z_i) \neq 0$ for all i, the zeros of f are all distinct, and in fact $z_{i+1} > 2z_i$ for all i.

We now list the first four zeros of f to 13 digits, as computed by Arnon Rosenthal.

$$z_0 = 1.488078545600$$
$$z_1 = 4.881140894897$$
$$z_2 = 13.56040852796$$
$$z_3 = 34.77531624775$$

(Furthermore, $z_4 \sim 85$ and we conjecture that $z_i = (i+1)2^i + o(2^i)$.)

For each n, let

$$f_n(x) = \prod_{i=0}^{n} (1-x/z_i).$$

In view of (15) and $z_{i+1} > 2z_i$,

$$\lim_{n\to\infty} f_n(x) = f(x)$$

for each x. A familiar result on partial fraction expansions, see for example Feller [3, p. 276], gives

$$\frac{1}{f_n(x)} = - \sum_{i=0}^{n} \frac{1}{z_i f_n'(z_i)(1-x/z_i)} .$$

Taking the limit in n, it follows that

$$(17) \qquad \frac{1}{f(x)} = - \sum_{i=0}^{\infty} \frac{1}{z_i f'(z_i)(1-x/z_i)}$$

is valid whenever x is not a zero of f. We may now expand both sides of (17) about 0 in power series with radii of convergence $\geq z_0$, obtaining

$$(18) \qquad \sum_{i=0}^{\infty} \frac{a_i x^i}{i!\, 2^{\binom{i}{2}}} = - \sum_{j=0}^{\infty} \frac{1}{z_j f'(z_j)} \sum_{i=0}^{\infty} \frac{x^i}{z_j^i} ,$$

in view of Corollary 1.

By comparing the coefficients of x^i on either side of (18) and applying the functional equation (16), it is seen that

$$(19) \qquad a_i = i!\, 2^{\binom{i}{2}} \sum_{j=0}^{\infty} \frac{1}{z_j f(z_j/2) z_j^i} .$$

To estimate the constants $1/z_j f(z_j/2)$, notice that from (15) we have

$$f'(z_j) = -\frac{1}{z_j} \prod_{i \neq j} \left(1 - \frac{z_j}{z_i}\right).$$

Using (16) and $z_{i+1} > 2z_i$, it follows that if we set $u = \prod_{n=1}^{\infty} (1-2^{-n})^{-1}$, then

$$0 < \frac{(-1)^j}{z_j f(z_j/2)} < \frac{u^2}{2^{\binom{j+1}{2}}} .$$

By this estimate, it is seen that for large i, it would take about $i(\sqrt{5} - 1)/2$ terms of the summation (19) to obtain a_i as the nearest integer.

Similarly Corollary 2 can be applied to find asymptotic estimates of $a_i^{(n)}$ for any fixed n. First, it is easy to see directly that

$$\Delta\left(\frac{x^n}{n!} e^{-x}\right) = \frac{x^n}{n!\, 2^{\binom{n}{2}}} f\left(\frac{x}{2^n}\right),$$

so Corollary 2 becomes

$$(20) \qquad \sum_{i=0}^{\infty} \frac{a_i^{(n)} x^i}{i!\, 2^{\binom{i}{2}}} = \frac{x^n f\left(\frac{x}{2^n}\right)}{n!\, 2^{\binom{n}{2}} f(x)} .$$

Then one finds that for all i,

$$(21) \quad a_i^{(n)} = \frac{i!\, 2^{\binom{i}{2}}}{n!\, 2^{\binom{n}{2}}} \sum_{j=0}^{\infty} \frac{z_j^{n-1} f(\frac{z_j}{2^n})}{f(\frac{z_j}{2})} \left(\frac{1}{z_j}\right)^i ,$$

the derivation of (21) being parallel to that of (19) by way of (17) and (18). For n = 1, (21) takes the particularly pleasing form

$$a_i^{(1)} = i!\, 2^{\binom{i}{2}} \sum_{j=0}^{\infty} \left(\frac{1}{z_j}\right)^i .$$

Setting

$$r_n = \frac{z_0^n\, f(\frac{z_0}{2^n})}{n!\, 2^{\binom{n}{2}}}$$

and using the first term of (19) and (21), we find that

$$a_i \sim \frac{i!\, 2^{\binom{i}{2}}}{r_1 z_0^i}$$

and for all $n \geq 1$,

$$a_i^{(n)} \sim r_n a_i .$$

In Table 2, we give the first few values of r_i and $\Sigma(j \leq i) r_j$, also as computed by Arnon Rosenthal.

It can be verified directly that $\sum_{j=1}^{\infty} r_j = 1$ by expanding f in a Taylor series about the point $x = z_0$, and evaluating it at 0. Obviously, the

Table 2

r_i and $\Sigma(j \le i) r_j$ for $1 \le i \le 8$

i	r_i	$\sum_{j \le i} r_j$
1	.574362373309	.574362373309
2	.366213673197	.940576046506
3	.056464543488	.997040589994
4	.002902307214	.999942897208
5	.000056651690	.999999548898
6	.000000449589	.999999998487
7	.000000001511	.999999999998
8	.000000000002	1.000000000000

r_i's are a probability distribution, because the chance that a random acyclic digraph has exactly i out-points is r_i.

Let K be the set of weakly connected acyclic digraphs. Of course, a digraph is acyclic if and only if every weak component is acyclic. Hence K(x) is determined from A(x) by Lemma 2, which gives

$$A(x) = e^{K(x)} . \tag{22}$$

We obtain

$$A'(x) = K'(x)A(x) \tag{23}$$

upon differentiating both sides of (22) by x. If k_n is the number of weakly connected acyclic digraphs on n points, (23) readily gives the recurrence relation

$$k_n = a_n - \sum_{s=1}^{n-1} \binom{n-1}{s-1} k_s a_{n-s} \tag{24}$$

for $n > 0$, with $k_0 = 0$. Gilbert [4] was the first to observe that (24) holds in a fairly general setting for labeled graphs; the connection between (22) and (24) was observed by Wright [18, p. 558].

If one chooses to read (23) as

$$K'(x) = \frac{A'(x)}{A(x)} , \tag{25}$$

then it is natural to let the generating function $g(x) = \Sigma(i\geq 1) g_i\ x^i/i!$ be determined by

$$A(x)(1 + g(x)) = 1. \tag{26}$$

From (26) and (25), we have the recurrence relations

$$g_n = -a_n - \sum_{s=1}^{n-1} \binom{n}{s} g_s\ a_{n-s},$$

$$k_n = a_n + \sum_{s=1}^{n-1} \binom{n-1}{s} g_s\ a_{n-s},$$

for all $n > 0$. The first few values are $g_1 = -1$, $g_2 = -1$, $g_3 = -13$, $g_4 = -373$, $g_5 = -24{,}061,\ldots$, $k_1 = 1$, $k_2 = 2$, $k_3 = 18$, $k_4 = 446$, $k_5 = 26{,}430,\ldots$

By Wright's Theorems 2 and 7 in [18], it follows that for $r \geq 1$,

$$k_n = a_n + \sum_{s=1}^{r-1} \binom{n}{s} g_s\ a_{n-s} + O(\binom{n}{r} a_{n-r}). \tag{27}$$

From (19) it is clear that for all $r,t \geq 0$,

$$\binom{n}{r} a_{n-r} = \frac{n!2^{\binom{n-r}{2}}}{r!} \sum_{i=1}^{t-1} \frac{1}{z_i^{n+1-r} f(z_i/2)} + a_n O((\frac{z_0}{z_t 2^r})^n) \tag{28}$$

From (27) and (28), one can find explicitly an approximation to k_n with error $a_n O(2^{-rn})$, for any r. For example

$$k_n = a_n(1 - \frac{4z_0}{2^n} + O(\frac{1}{4^n})).$$

Let $K^{(m)}$ be the set of weak acyclic digraphs with exactly m out-points. It is easy to see that

$$K^{(1)}(x) = A^{(1)}(x),$$

$$K^{(2)}(x) = A^{(2)}(x) - \frac{1}{2} A^{(1)}(x)^2,$$

$$K^{(3)}(x) = A^{(3)}(x) - \frac{1}{2}A^{(1)}(x)K^{(2)}(x) - \frac{1}{6} A^{(1)}(x)^3,$$

and so forth. This sequence of equations is summarized by the single equation

$$(29) \qquad e^{\sum_{i=1}^{\infty} y^i K^{(i)}(x)} = \sum_{j=0}^{\infty} y^j A^{(j)}(x).$$

If $k_n^{(i)}$ is the number of digraphs in $K^{(i)}$ with exactly n points, then for fixed i an asymptotic expansion for $k_n^{(i)}$ can be obtained from (29) and (21).

4. Digraphs which are strong, have a source, are unilateral, or are weak.

Let D be the set of all digraphs and S the set of all strongly connected digraphs (both labeled, as usual). Then

$$(30) \qquad \Delta D(x) = \sum_{i=0}^{\infty} \frac{2^{\binom{i}{2}} x^i}{i!}$$

since there are $d_i = 2^{i(i-1)}$ digraphs on i points. By Theorem 1,

(31) $$\Delta D(x) = (\Delta e^{-S(x)})^{-1}$$

To find recurrence relations for the number s_n of strongly connected digraphs on n points, we introduce the series $h(x) = \Sigma(i\geq 1)h_i x^i/i!$ such that

(32) $$\Delta h(x) = 1 - \frac{1}{\Delta D(x)} \quad .$$

Now (30) and (32) give the relation

(33) $$h_n = 2^{n(n-1)} - \sum_{t=1}^{n-1} \binom{n}{t} 2^{t(n-1)} h_{n-t}$$

for all $n > 0$. On the other hand (31) and (32) give

$$1 - \Delta h(x) = \Delta e^{-S(x)}, \text{ or}$$

$$S(x) = -\ell n(1-h(x)).$$

Differentiating with respect to x, we find

$$S'(x) = \frac{h'(x)}{1-h(x)}, \text{ or}$$

$$S'(x) = h'(x) + h(x)S'(x).$$

Equating coefficients of x^n is easily seen to yield

(34) $$s_n = h_n + \sum_{t=1}^{n-1} \binom{n-1}{t} s_{n-t} h_t .$$

The recurrence relations (33) and (34) are the ones

which Wright [19] derived from a more complicated set of recurrence relations for s_n due to Liskovec [12].

Let $D^{(n)}$ be the set of all digraphs which have exactly n out-components. Theorems 1 and 2 give

$$\Delta D^{(n)}(x) = \Delta D(x)\ \Delta(\frac{S(x)^n}{n!}\ e^{-S(x)}),$$

and in particular

$$\Delta D^{(1)}(x) = \Delta D(x)\Delta(S(x)\ell^{-S(x)}). \tag{35}$$

A point v of a digraph is said to be a <u>source</u> just if every other point of the digraph can be reached from v. A necessary and sufficient condition for v to be a source is that the digraph contain exactly one out-component and that v lie in it. Consequently, the number $d_n^{(1)}$ of digraphs on n points in $D^{(1)}$ is also the number of digraphs on n points which have a source.

We proceed to put (35) in the form of a pair of recurrence relations for $d_n^{(1)}$. From (31) and (32) it follows that

$$e^{-S(x)} = 1 - h(x).$$

Let the series $\Sigma(j\geq 1)q_j x^j/j!$ be the product

$S(x)e^{-S(x)}$. Then for all $j > 0$,

(36) $$q_j = s_j - \sum_{t=1}^{j-1} \binom{j}{t} s_{j-t} h_t .$$

Moreover (35) becomes

$$\Delta D^{(1)}(x) = \sum_{i=0}^{\infty} \frac{2^{\binom{i}{2}} x^i}{i!} \sum_{j=1}^{\infty} \frac{q_j x^j}{j! 2^{\binom{j}{2}}} ,$$

which is quickly seen to give the relation

(37) $$d_n^{(1)} = q_n + \sum_{i=1}^{n-1} \binom{n}{i} 2^{i(n-1)} q_{n-i}$$

for all $n > 0$.

A digraph δ is <u>unilaterally connected</u> (or <u>unilateral</u>) if for any two points of δ, at least one is reachable from the other. An obvious necessary and sufficient condition for δ to be unilateral is that δ have exactly one out-component, with the subgraph induced by the other strong components of δ being in turn unilateral. This characterization of unilateral digraphs, with the help of observations in Section 2, leads to the double recurrence relations for the numbers $u_{p,k}$ of unilateral digraphs with p points in the out-component and $p + k$ points in all:

$$(38) \qquad u_{p,k} = \binom{p+k}{k} s_p \sum_{m=1}^{k} (2^{pk} - 2^{p(k-m)} u_{m,k-m},$$

which is valid if p,k > 0, and

$$(39) \qquad u_{p,0} = s_p,$$

which is valid if p > 0. The relations (38) and (39) determine $u_{p,k}$ from the numbers s_p, by induction on k. Of course, the total number u_n of unilateral digraphs on n points is

$$(40) \qquad u_n = \sum_{m=1}^{n} u_{m,n-m} .$$

In Table 3, we present the values of s_n, u_n, and $d_n^{(1)}$ for $n \leq 8$, which were calculated using equations (33)-(40). The values of s_n for $n \leq 6$ were computed by Liskovec [12].

For comparison we also include in Table 3 the numbers d_n and w_n, the latter being the number of weak digraphs on n points. The numbers w_n were found by applying Lemma 2. This leads to the relation

$$(41) \qquad w_n = 2^{n(n-1)} - \sum_{t=1}^{n-1} \binom{n-1}{t} 2^{t(t-1)} w_{n-t},$$

valid for all $n \geq 1$, in the same way that (24) was obtained in Section 3.

Table 3

The values of $d_n, w_n, d_n^{(1)}, u_n$, and s_n for $n \leq 8$.

n	d_n	w_n	$d_n^{(1)}$	u_n	s_n
0	1	0	0	0	0
1	1	1	1	1	1
2	4	3	3	3	1
3	64	54	51	48	18
4	4,096	3,834	3,614	3,400	1,606
5	1,048,576	1,027,080	991,930	955,860	565,080
6	1,073,741,824	1,067,245,748	1,051,469,032	1,034,141,596	734,774,776
7	4,398, 046,511,104	4,390, 480,560,744	4,364, 841,320,040	4,338, 541,672,792	3,520, 944,131,920
8	72,057,594, 037,927,936	72,022,346, 390,883,864	71,943,752, 944,978,224	71,839,019, 692,720,536	63,569,709, 615,130,544

5. Related problems.

The methods of the present paper can be combined with those of [15] to count unlabeled digraphs with given strong components. This leads to methods for counting acyclic, strong, and unilateral digraphs, as well as digraphs with a source, all unlabeled. Similarly, unlabeled strong tournaments can be counted; the labeled ones were counted by Liskovec [12].

Extensions of the methods of [15] make it possible to count unlabeled self-converse digraphs which are acyclic, strong, unilateral, or have a source, and also unlabeled self-converse strong tournaments. These results will appear elsewhere.

A puzzling unsolved problem is to count self-converse labeled digraphs. Unlabeled self-converse digraphs were counted easily by Harary and Palmer [8]. It is very rare that a labeled counting problem is harder than the corresponding unlabeled problem.

The enumeration of transitive digraphs, labeled or not, is a problem of long standing intractibility.

Also open is the problem of counting strong digraphs which have strong complements. Strong digraphs with weak complements are easily counted, for any non-empty digraph which is not weak has a strong complement, and these may be subtracted from the total number of strong digraphs. Another open problem is to count strong digraphs which are self-complementary.

REFERENCES

[1] E. A. Bender and J. R. Goldman, Enumerative uses of generating functions. Indiana Univ. Math. J. 20 (1971), 753-765.

[2] C. C. Cadogan, The möbius function and connected graphs. J. Combinatorial Theory 11B (1971), 193-200.

[3] W. Feller, An Introduction to Probability Theory and Its Applications, Vol. I, Third Edition, Wiley, New York, 1968.

[4] E. N. Gilbert, Enumeration of labelled graphs. Canad. J. Math. 8 (1956), 405-411.

[5] J. S. Hadamard, Étude sur les proprietés des fonctions entières et en particulier d'une fonction considérée par Riemann. J. Math. (4)9 (1893), 171-215.

[6] F. Harary, Unsolved problems in the enumeration of graphs, Publ. Math. Inst. Hungar. Acad. Sci. 5 (1960), 63-95.

[7] F. Harary, R. Z. Norman and D. Cartwright, Structural Models: An introduction to the theory of directed graphs, Wiley, New York, 1965.

[8] F. Harary and E. M. Palmer, Enumeration of self-converse digraphs. Mathematika 13 (1966), 151-157.

[9] F. Harary and E. M. Palmer, Graphical Enumeration. Academic Press, New York, 1973, to appear.

[10] F. Harary and R. C. Read, Is the null graph a pointless concept?, to appear.

[11] E. N. Laguerre, Mémoire sur la théorie des équations numériques. J. Math. (3) 9 (1883), 99-146.

[12] A. Liskovec, On a recurrence method of counting graphs with labelled vertices. Soviet Math. Dokl. 10 (1969), 242-256. (Dokl. Akad. Nauk SSSR, 184 (1969), No. 6.)

[13] K. Mahler, On a special functional equation. J. London Math. Soc. 15 (1940), 115-123.

[14] G. Pólya and J. Schur, Uber zwei Arten von Faktorenfolgen in der Theorie der algebraischen Gleichungen. J. Reine Angew. Math. 144 (1914), 89-113.

[15] R. W. Robinson, Enumeration of non-separable graphs. J. Combinatorial Theory 9 (1970), 327-356.

[16] R. W. Robinson, Enumeration of acyclic digraphs. Combinatorial Mathematics and Its Applications (R. C. Bose et al. eds.). Univ. of North Carolina, Chapel Hill, 1970. pp. 391-399.

[17] G. E. Uhlenbeck and G. W. Ford, Lectures in Statistical Mechanics. American Math. Soc., Providence, R.I., 1963.

[18] E. M. Wright, Asymptotic relations between enumerative functions in graph theory. Proc. London Math. Soc. 20 (1970), 558-572.

[19] E. M. Wright, The number of strong digraphs. Bull. London Math. Soc. 3 (1971), 348-350.

ALMOST ALL TREES ARE COSPECTRAL

Allen J. Schwenk

Abstract. We show that for p sufficiently large, almost every tree T with p points has a cospectral mate, that is, a second tree which has the same characteristic polynomial as T. To obtain this result, we define a limb of T to be a certain type of rooted subtree of T. We then observe that there happen to be two limbs with 8 lines such that if T has one of these limbs, then it may be replaced by the other limb without changing the spectrum of T. Thus, any tree with one of these two particular limbs has a cospectral mate. The bulk of our effort is required to show that almost every tree has one of these two limbs.

Curiously, we find that the number of trees having any specified limb depends solely upon the size of that limb and not upon its structure. Knowing this, we are able to enumerate the trees with any forbidden limb of weight n, and so, by complementation, we count the number of trees with that specified limb of weight n. An asymptotic analysis confirms that given any specified limb, almost every tree has that limb. Applying this result to one of our 8-line limbs, we obtain the main result.

In addition, this asymptotic analysis provides an independent proof of two known results, namely that almost all trees have a nontrivial automorphism, and almost all trees are homeomorphically reducible.

1. Introduction.

When the spectrum of a graph was first studied, it was hoped that it would characterize the graph. Counterexamples were quickly found, for trees as well as other graphs, so restrictions were added, still hoping that some nontrivial subclass of graphs could be characterized by their spectra.

The earliest known cospectral pairs were generally found by exhaustive search and much duplication of results occurred. The following account is not intended to assign credit for various discoveries, but rather to suggest that it is unlikely that *any* list of additional restrictions will produce a nontrivial subclass of graphs in which each graph is determined by its spectrum.

The smallest pair of cospectral graphs has 5 points [7], but one of these is not connected. The smallest connected pair has 6 points [1,2,7]. Fisher [4] was the first to find a pair of cospectral blocks, which have 15 points, and Baker [1] found the smallest pair of cospectral blocks with 7 points. Hoffman [10] constructed a cospectral pair of regular graphs in which one of the graphs was Q_4, the four dimensional cube, and Shrikhande [14] even found a cospectral pair of strongly regular graphs, also with 16 points. Furthermore, cospectral triples [1,2,7] and even a quadruple [1] have been discovered. The smallest pair of cospectral trees have 8 points [1,2,3,7], and even an isomeric pair has been presented [2]. In fact, the smallest homeomorphic pair with 9 points has been displayed [7].

Still, chemists [15], who wanted to use the spectrum as a classification scheme, and others (see [7]) conjectured that graphs with cospectral mates were the exception and not the rule. In this paper, we show that, at least for trees, this conjecture is as wrong as possible. Namely, asymptotically, almost every tree has a cospectral mate.

The proof of this result is rather involved, and so we outline it in this section. In Section 2, we present a characterization which provides a convenient test of whether two graphs are cospectral, and we specialize this to trees. We also present a result which expresses the characteristic polynomial of a tree T in terms of certain subtrees which we call limbs. We use these results to show that there is a certain pair of 9-point rooted trees R and S such that if T has R as a limb, then the limb can be replaced by S to form a new tree which is cospectral with T. Thus, our main result will follow if we can simply show that almost every tree has R as a limb.

We find in Section 3 that the number of trees on p points <u>not</u> having a specified limb is independent of the structure of the limb but depends solely upon the weight of the limb. Knowing this, we can count trees with a forbidden limb of weight n.

We proceed in Section 4 to examine the growth rate of the series found in Section 3 and, in particular, to show that it is strictly less than the growth rate for ordinary trees. Thus, asymptotically the fraction of trees with a forbidden limb as compared to all trees goes to zero. Consequently, we conclude that given any rooted tree R, almost every tree has R as a limb.

Finally, in Section 5, we formally state the main theorem which is then an immediate consequence of the results already obtained. We conclude by commenting on the likelihood of the conjecture that almost all *graphs* have a cospectral mate.

2. A method for constructing cospectral trees.

Let $\phi(G;x) = \phi(G) = \sum_{n=0}^{p} a_n x^{p-n}$ denote the characteristic polynomial of G, and let c(G) and k(G) be the numbers of cycles and components in G. In a *mutation graph*, each component is either a cycle or K_2; these graphs were introduced in Harary [5]. Each permutation with no fixed points can be viewed as a one-to-one mapping taking each point of the underlying mutation graph to one of its neighbors. Note that a given mutation graph G corresponds to $2^{c(G)}$ permutations, since each cycle may be mapped in either of two directions. Finally, let G_n be the set of those n-point subgraphs of G which are mutations. Using this notation, we may concisely state a result of Sachs [13].

Theorem 1. For $n \geq 1$, $a_n = \sum_{H \in G_n} (-1)^{k(H)} (2)^{c(H)}$.

If we restrict the graph to be a tree T, then the mutation subgraphs must be sets of independent lines since T contains no cycles. Thus, only even ordered mutations can exist for trees.

Corollary 1. If T is a tree and $n \geq 1$, then

$$a_{2n-1} = 0 \text{ and}$$
$$a_{2n} = \sum_{H \varepsilon T_{2n}} (-1)^n = (-1)^n |T_{2n}|.$$

In other words, only the even subscripts appear, and these count the number of ways to select n independent lines from T, that is, n lines with no pair incident.

As in Harary [6, p. 35], we define a branch of a tree at the point v to be a maximal subtree containing v as an endpoint. The union of one or more branches at v is called a limb at v which we view as being rooted at v. The weight of a limb is the number of lines in it. These concepts are illustrated in Figure 1.

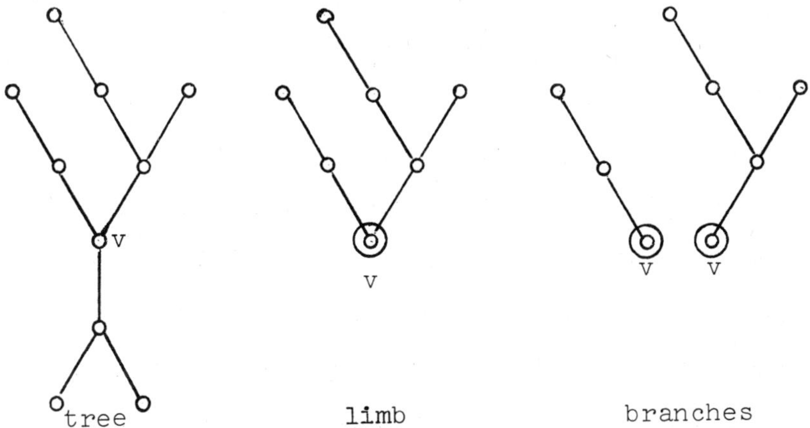

Figure 1. A tree with a limb of weight 6 consisting of two branches.

The rooted tree formed by identifying the roots of two rooted trees L_1 and L_2 is denoted $L_1 \cdot L_2$ and is called the coalescence of L_1 and L_2. Thus, this tree may be viewed as having two line-disjoint limbs L_1 and L_2 at its root v.

Theorem 2. The characteristic polynomial of the coalescence of two rooted trees is given by:

$$\phi(L_1 \cdot L_2) = \phi(L_1 - v)\phi(L_2) + \phi(L_1)\phi(L_2 - v) - x\phi(L_1 - v)\phi(L_2 - v).$$

Proof. According to Corollary 1, we need only show that each set of n independent lines is counted precisely once on the right hand side in the coefficient of $(-1)^n x^{p-2n}$.

Now the set of n lines contains at most one line incident with v. Thus, we observe that if the set contains no line incident with v in L_1, then it is counted by the first term since its lines lie in $L_2 \cup (L_1 - v)$. Similarly, if it has no lines incident with v in L_2, then it is counted by the second term. However, those sets having no line of either limb incident with v have now been counted twice, so we subtract the third term to correct for this. The factor of x is needed because the single point v has been removed twice, once from each limb.

Let R and S be the limbs shown in Figure 2 which result from two different rootings of the same tree. We claim that any tree with limb R is cospectral to the tree obtained when R is replaced by S. Equivalently, we claim that, for any rooted tree T, the two coalescences $T \cdot R$ and $T \cdot S$ are cospectral.

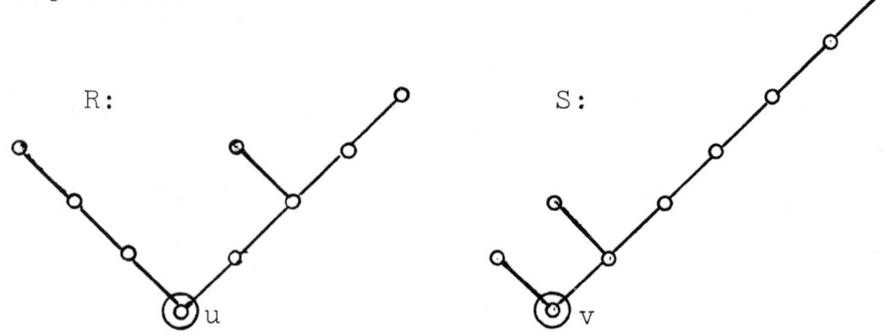

Figure 2. Two limbs of weight 8.

<u>Theorem 3</u>. For any rooted tree T and for R and S shown in Figure 2,

$$\phi(T \cdot R) = \phi(T \cdot S).$$

<u>Proof</u>. If we expand each side of the above equation by Theorem 2, we see that it suffices to show that $\phi(R) = \phi(S)$ and $\phi(R-u) = \phi(S-v)$. But the first of these two equations is trivial because $R = S$ if we ignore the labels on the roots. For the second equation we see that

(1) $\phi(R-u) = \phi(\;\;)\phi(P_3) = (x^5-4x^3+2x)(x^3-2x).$

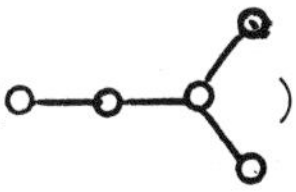

Similarly,

(2) $\phi(S-v) = \phi(K_1)\phi(P_7) = x(x^7-6x^5+10x^3-4x),$

and we see that $\phi(R-u) = \phi(S-v)$ as required.

In fact, we can proceed to produce large collections of cospectral trees by taking a tree with many limbs isomorphic to R. Then each replacement of a copy of R by a copy of S produces a tree cospectral to the original tree. But our goal is to show that almost every tree has R as a limb, and so has at least one cospectral mate which can be obtained by replacing limb R by a limb S. Thus, we must count the trees not having R as a limb in order to show that, asymptotically, the fraction of these trees divided by all trees goes to zero. The next two sections prove this result for any limb.

3. Trees with forbidden limbs.

As usual, we let $t(x) = \sum_{p=1}^{\infty} t_p x^p$ and $T(x) = \sum_{p=1}^{\infty} T_p x^p$ denote the generating functions for trees and rooted trees.

Theorem 4. If R and S are two rooted trees of weight n, and r_p (s_p) is the number of p-point trees which do not have R (S) as a limb, then $r_p = s_p$.

Proof. There are three cases.

Case 1. $p \leq n$.

In this case, it is trivial that a p-point tree has neither R nor S as a limb since it has only $p - 1 < n$ lines. Thus $r_p = s_p = t_p$.

Case 2. $n + 1 \leq p \leq 2n$.

It suffices to show that $t_p - r_p = t_p - s_p$, in other words, as many trees have R as a limb as have S. Now let each tree T having R as a limb inherit the root of its R. We claim that this produces a unique rooting of T, even if T has more than one copy of R, because in that instance, all the roots of these copies of R lie in the same similarity class of T. We state this needed result formally:

Lemma 4. If L is a limb of T of weight n with $p \leq 2n$, and L_1, L_2 are occurrences of L at v_1, v_2, then T has an automorphism α interchanging roots v_1 and v_2 and interchanging limbs L_1 and L_2.

Outline of Proof. The proof of this lemma in all its devilish detail is terribly tedious, so we shall just provide an outline. We use a two step induction on the distance $d(v_1, v_2)$. If $d(v_1, v_2) = 0$, we simply interchange the branches of L_1 with the corresponding branches of L_2. If $d(v_1, v_2) = 1$, inspection confirms that for each branch at v_1 there is a corresponding branch at v_2. We choose an automorphism which interchanges these branches. Now let $d(v_1, v_2) \geq 2$ and assume the lemma has been proved for distances less than $d(v_1, v_2)$.

Let v_1' and v_2' lie on the path joining v_1 and v_2 with $d(v_1, v_1') = d(v_2, v_2') = 1$. Moreover, let L_1'

be the largest limb of L_1 at v_1' not containing v_1. Define L_2' accordingly. Now v_1', v_2', L_1', and L_2' satisfy the hypothesis of the lemma, and $d(v_1', v_2') = d(v_1, v_2) - 2$. Thus, by the induction hypothesis, there is an automorphism α interchanging v_1' and v_2'. We inspect α and see that it happens to interchange v_1 and v_2 and also L_1 and L_2, completing the proof.

We return to the proof of the theorem. We have shown that if T has limb R, then a unique rooting of T is provided by the root of R. But clearly such rooted trees are in one-to-one correspondence with arbitrary rooted trees on p-n points, because $T = R \cdot T^*$ where T^* is a rooted tree with p-n points. In other words, we have shown that $t_p - r_p = T_{p-n}$. But the same argument applies to S, so
$t_p - s_p = T_{p-n} = t_p - r_p$.

Case 3. $p \geq 2n + 1$.

In this case, we wish to establish a one-to-one correspondence between those trees having R as a limb but not S, and those having S but not R. Our idea is simply to replace each occurrence of R by an occurrence of S until no more limbs R remain. Thus, we convert a tree T having R as a limb but not S to a tree T* having S but not R. This does indeed work, but care is needed to justify the procedure. We first note that two limbs of weight n at different points v_1 and v_2 cannot intersect, for if they do, the intersection must contain the path joining v_1 and v_2. Consequently, every line of T must lie in at least one of the two limbs, and since those lines on the path joining v_1 and v_2 lie in both limbs we have $2n \geq p - 1 + d(v_1,v_2) \geq p$ in violation of the hypothesis of this case. Thus, when we replace R by S at v_1, we neither create nor destroy any limb R or S rooted at another point. This permits us to focus our attention at one point v_1. The procedure used is then applied successively to each point of T.

The temptation is to say "Let k be the maximum number of line-disjoint occurrences of R at v_1. Then replace these by k line-disjoint copies of S." This procedure may fail to remove all occurrences of R at v_1 if R and S happen to have a common branch (see Figure 3). This procedure may be salvaged by continuing to remove R's. Let us determine just how many copies of R must be replaced by limbs S.

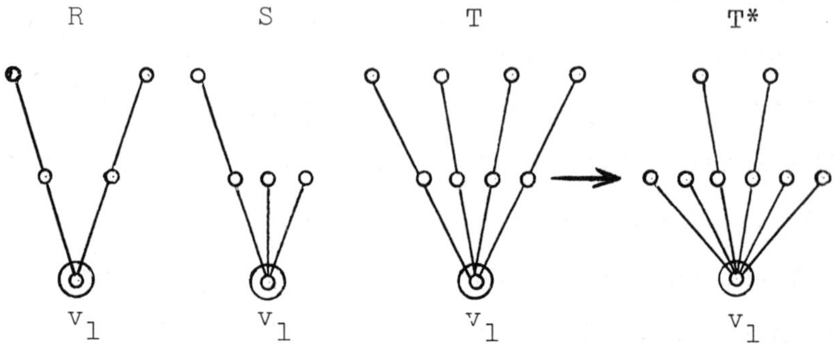

Figure 3. Failure of the intuitive procedure.

Let $B_1, B_2, \ldots, B_m$ be the collection of rooted branches of weight at most n. We may write $R = \bigcup_{i=1}^{m} a_i B_i$ to mean R has exactly a_i copies of branch B_i at v_i. Similarly, let $S = \bigcup_{i=1}^{m} b_i B_i$ and let $T = \bigcup_{i=1}^{m} c_i B_i$. Finally, let $J = \{i : a_i < b_i\}$ and $K = \{i : a_i > b_i\}$. Note that J and K are not empty since $R \neq S$ implies that R has some branch not in S and S has some branch not in R. Let

$$N = \min \{[\frac{c_i - b_i}{a_i - b_i}] : i \in K\}. \tag{3}$$

We claim that N is the number of times R must be replaced by S at v_1 in order to remove every occurrence of R at v_1. To see this, examine that particular $k \in K$ which attains the minimum. Then

$$(4) \qquad 0 \leq c_k - b_k - N(a_k - b_k) < a_k - b_k,$$

and so

$$(5) \qquad b_k \leq c_k - N(a_k - b_k) < a_k.$$

Thus, after N removals of R, each of which consumes $a_k - b_k$ copies of branch B_k, there remain less than a_k copies of B_k so we cannot remove another copy of R. On the other hand, after N - 1 removals of R, there remain

$$(6) \qquad c_i - (N - 1)(a_i - b_i) \geq a_i$$

branches B_i for each $i \in K$, so one more R can be removed. Let the tree T* so formed have d_i branches B_i where $d_i = c_i - N(a_i - b_i)$. Now consider those $i \in J$. Since T doesn't have S as a limb at v_1, there is some $j \in J$ such that $b_j > c_j$. For this j, $d_j = c_j + N(b_j - a_j)$ and so

$$(7) \quad \min \{[\frac{d_i - a_i}{b_i - a_i}] : i \in J\} = [\frac{d_j - c_j}{b_j - a_j}] = N$$

since $d_j - N(b_j - a_j) = c_j$ which is positive yet less than b_j.

The argument is complete. If we start with a tree T which has R as a limb but not S, we proceed to remove R's until we form a tree T* having S as a limb but not R. Reversing the procedure, we return to the original tree T. Thus the set of trees having R but not S as a limb is in one-to-one correspondence with the set of trees having S but not R. Thus $r_p = s_p$ completing the proof of Theorem 1.

Any proof that requires inspection of arbitrary cases is unappealing. Case 1, could have been subsumed vacuously under either of the other two cases, but it was isolated for clarity as the only easy case. The other two cases represent a very real and apparently necessary division. The distinction is that in Case 2, two copies of a limb at different points must overlap while in Case 3, they must be disjoint. Because of this difference, it seems that neither argument can be extended to cover the other case.

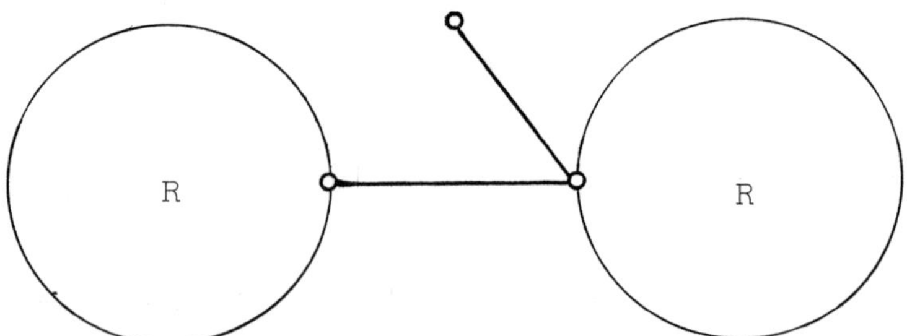

Figure 4. A tree with two nonisomorphic rootings at roots of limb R.

For example, the first argument requires T to be rooted uniquely at the root point of its R. Figure 4 depicts a tree with 2n + 3 points having two nonisomorphic possible rootings at roots of R. On the other hand, the second argument depends upon replacing R's by S's until no more R's remain. If we try to apply this procedure when $p \leq 2n$, it need not terminate as suggested by Figure 5.

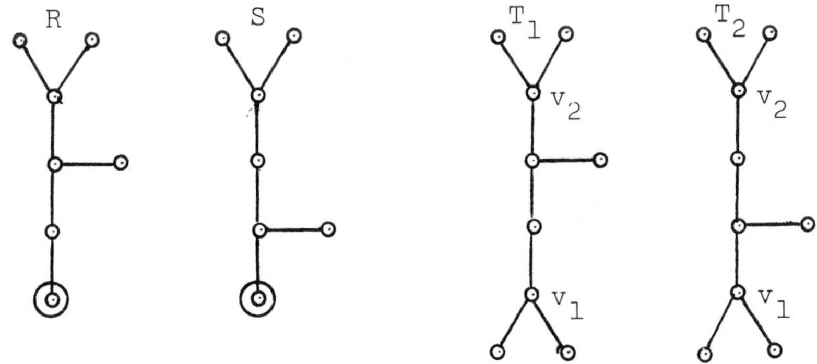

Figure 5. A tree (from Case 2) for which the procedure of Case 3 never terminates.

Tree T_1 has R at v_1, so we replace it by S and obtain T_2 which, in turn, has R at v_2. Replacing this R by S returns us to T_1. Admittedly, it might be possible to patch the proof since the tree of Figure 5 is not one to which the procedure need be applied, yet this possibility raises serious doubts as to whether the argument could be salvaged. Here the necessary overlapping of two copies of the same limb produces complicating alterations as one tries to carry out the described procedure. We are left with the opinion that two distinct cases are necessary.

Nevertheless, we have shown that the number of trees not having a forbidden limb of weight n is independent of the structure of the limb. Thus we let

$$r^{(n)}(x) = \sum_{p=1}^{\infty} r_p^{(n)}(x) \tag{8}$$

be the generating function counting trees not having a specified limb of weight n, and we take the star $K_{1,n}$ rooted at its center to be the canonical example of a forbidden limb. Let $R^{(n)}(x)$ denote the series for rooted trees not having $K_{1,n}$ as a limb and let $Q^{(n)}(x)$ denote the series for rooted trees with exactly n endlines at the root and no other occurrence of $K_{1,n}$ as a limb. Furthermore, let $\overline{Q}^{(n)}(x)$ denote the same algebraic series, only this time considered as representing these same trees with the root moved to one of the n equivalent endpoints mentioned above. Finally, let

$$S^{(n)}(x) = Q^{(n)}(x) + R^{(n)}(x) = \overline{Q}^{(n)}(x) + R^{(n)}(x). \tag{9}$$

Theorem 5. The generating function for the number of trees having a forbidden limb of weight n is given by the two equations

$$(10)\quad r^{(n)}(x) = S^{(n)}(x) - \tfrac{1}{2}[(S^{(n)}(x))^2 - S^{(n)}(x^2)]$$

$$(11)\quad S^{(n)}(x) = (x - x^{n+1}) \prod_{r=1}^{\infty} (1-x^r)^{-S_r^{(n)}} .$$

Alternatively, $S^{(n)}(x)$ satisfies the functional equation

$$(12)\quad S^{(n)}(x) = (x - x^{n+1}) \exp \sum_{k=1}^{\infty} \frac{1}{k} S^{(n)}(x^k).$$

Proof. Throughout this proof we deal with only one value of n, and so we shall suppress the superscript n on all counting series and write r(x), Q(x), R(x), and S(x).

We start by finding an expression for $R(x) + \overline{Q}(x)$, the number of rooted trees T either having no $K_{1,n}$ as a limb or having just one, and the root being at one of the n equivalent endpoints. We let the neighbors of the root v of T inherit the root, and delete v to form a rooted forest.

What kinds of trees can appear in this forest? If T is counted by $\overline{Q}(x)$, then deleting v destroys the occurrence of $K_{1,n}$ as a limb, and so produces a tree counted by R(x). If, on the other hand, T is counted by R(x), each component of T - v must have no $K_{1,n}$ as a limb which would survive the process of rebuilding T. In other words, each component is either a tree counted by R(x), which has no $K_{1,n}$, or it is a tree having just one $K_{1,n}$ which is destroyed upon rejoining the roots of each component. This latter possibility happens exactly when T is counted by $\overline{Q}(x)$. Finally, T - v can contain no more than n - 1 trivial trees, for otherwise T would have $K_{1,n}$ as a limb at v, the root of T. Thus we have the formula

$$(13)\quad R(x)+\overline{Q}(x)=x(1+x+\ldots+x^{n-1})\prod_{r=2}^{\infty}(1+x^{r}+x^{2r}+\ldots)^{R_r+\overline{Q}_r}$$

where the first factor x accounts for the root, the factor $(1+x+ \ldots + x^{n-1})$ counts the number of trivial trees in the forest, and the remaining product counts the number of occurrences of each rooted tree from R and $\overline{Q}$ in the forest. Substituting $S(x) = R(x) + \overline{Q}(x)$ and applying routine algebraic manipulations, we obtain equation (11).

To convert this to equation (12), we take the logarithm of equation (11) to get

$$(14)\quad \log S(x) = \log(x-x^{n+1}) - \sum_{r=1}^{\infty} S_r \log (1-x^r).$$

Replacing log $(1-x^r)$ by its series, we get

$$(15)\quad \log S(x) = \log(x-x^{n+1}) + \sum_{r=1}^{\infty}\sum_{k=1}^{\infty} S_r \frac{x^{rk}}{k} .$$

Now the double sum is absolutely convergent, so we interchange the order of summation, and, noting that $\sum_{r=1}^{\infty} S_r x^{rk} = S(x^k)$, we obtain

$$(16)\quad \log S(x) = \log(x-x^{n+1}) + \sum_{k=1}^{\infty} \frac{1}{k} S(x^k).$$

Exponentiation now produces the functional equation (12).

Now in order to remove the rooting from R(x) to obtain r(x), we recall the dissimilarity characteristic equation for trees (see Harary [6 , p. 189]). Namely, if p^* and q^* are the number of similarity classes of points and lines, respectively, in a tree T, and s is the number of symmetry lines, then $s = 0$ or 1 and $p^* - (q^* - s) = 1$. Thus, if we add up these terms for each tree contributing to r(x), the net contribution will be 1 for each tree. Thus

$$(17)\qquad r(x) = \sum p^* - \sum q^* + \sum s.$$

But now for each tree counted in r(x), p^* gives the number of rootings it has, so $\Sigma p^* = R(x)$, the number of rooted trees not having $K_{1,n}$. Similarly, $\Sigma(q^*-s)$ counts the number of line-rooted trees, which we may view as the result of joining two distinct rooted trees with a line between their roots. Noting that these distinct trees may be from $R(x)$ or from $\overline{Q}(x)$ (since the added line will then destroy the single occurrence of $K_{1,n}$), we find that the correction term is the same as in Harary [6, p. 190, eq. (15.44)] with T replaced by $S = R + \overline{Q}$. We now have the expression

$$(18) \qquad R(x) - \frac{1}{2}\,(S^2(x) - S(x^2)),$$

and we have explicitly shown that each tree not having $K_{1,n}$ has been counted just once, but what about trees which <u>do</u> contain $K_{1,n}$? Do they contribute 0, as they should, to expression (18)? Those having more than one $K_{1,n}$ are not counted in any of the three terms $R,(x)$, $\frac{1}{2}S^2(x)$, and $\frac{1}{2}S(x^2)$. However, those in $\overline{Q}(x)$ have been inadvertently subtracted in the term $\frac{1}{2}S^2(x)$, and so to nullify this negative contribution, we add $\overline{Q}(x)$ to obtain

$$(19) \qquad r(x) = R(x) - \frac{1}{2}(S^2(x) - S(x^2)) + \overline{Q}(x).$$

This neatly simplifies to give equation (10) and completes the proof.

Those familiar with tree-counting will recognize equation (10) as Otter's formula for trees in terms of rooted trees. Equation (11) is analogous to Cayley's recursive expression for counting rooted trees, and equation (12) corresponds to Pólya's functional version of Cayley's result. (See Harary [6, pp. 187-190].) In fact, if we imagine taking a "limit" as n approaches infinity, we conceive of forbidding successively only larger stars until, in the limit, no star is forbidden. The problem then reverts to counting ordinary trees and rooted trees. In the "limit", equations (11) and (12) become Cayley's and Pólya's equations respectively, and so our intuition has been heuristically supported. Consequently, we shall view $t(x)$ as $r^{(\infty)}(x)$ and $T(x)$ as $S^{(\infty)}(x)$.

Equations (10) and (11) are readily usable by a computer program to generate the first several coefficients of $r^{(n)}(x)$. This has been done, and the first 39 coefficients of $r^{(n)}(x)$ have been computed for $2 \leq n \leq 9$. For comparison, the numbers of ordinary trees (with no forbidden limb) have also been generated up to $p = 39$.

4. Asymptotic Estimate.

We must now estimate the growth rate of $r^{(n)}(x)$ to assure that asymptotically the fraction of trees avoiding a specified forbidden limb of weight n goes to zero. This growth rate is just the reciprocal of the radius of convergence of the power series, and so our goal is to show that $r^{(n)}(x)$ converges on a larger disk than $t(x)$. Thus, we let $\eta_{(n)}$ and η denote the corresponding radii of convergence, and we proceed to show $\eta_{(n)} > \eta$.

In general, estimating the radius of convergence of a power series accurately is a very difficult problem. But in this case, we are fortunate because $r^{(n)}(x)$ and $t(x)$ are closely related and the radius of $t(x)$ has already been estimated by Pólya [12] and Otter [11]. Thus, we may estimate $\eta_{(n)}$ by imitating Otter's method for estimating η as presented in Harary and Palmer [8, Section 9.5], and we shall omit those tedious details which simply duplicate the material in that book.

Theorem 6. The radius of convergence for $r^{(n)}(x)$ is larger than the radius for $t(x)$.

Proof. We have observed that $r^{(n)}(x)$ is related to $S^{(n)}(x)$ by equation (10) in just the same way as $t(x)$ is related to $T(x)$. Moreover, this relation guarantees that $r^{(n)}(x)$ and $S^{(n)}(x)$ have the same radius of convergence, just as $t(x)$ and $T(x)$ converge on the same disk. Thus, we focus our attention on $S^{(n)}(x)$. Now since each coefficient satisfies $0 \leq S_p^{(n)} \leq T_p$, we know that $\eta_{(n)} \geq \eta$. Thus, it remains to show that $\eta_{(n)} \neq \eta$. This is done with the help of two intermediate results.

Lemma 6a. The limit of $S^{(n)}(x)$ as x approaches $\eta_{(n)}$ from the left exists and is given by

$$(20) \quad \lim_{x \to \eta_{(n)}^-} S^{(n)}(x) = S^{(n)}(\eta_{(n)}) = \sum_{p=1}^{\infty} S_p^{(n)} \eta_{(n)}^p .$$

Proof of Lemma. From equation (12) we see that

$$(21) \quad \frac{S^{(n)}(x)}{x} = (1 - x^n) \exp \sum_{k=1}^{\infty} \frac{1}{k} S^{(n)}(x^k),$$

which, upon taking logarithms becomes

$$(22) \quad \log(S^{(n)}(x)/x) = \log(1-x^n) + \sum_{k=1}^{\infty} \frac{1}{k} S^{(n)}(x^k).$$

We replace the $\log(1-x^n)$ by its power series to obtain

(23) $$\log(S^{(n)}(x)/x) = S^{(n)}(x) - \sum_{k=1}^{\infty} \frac{x^{kn}}{k} + \sum_{k=2}^{\infty} \frac{1}{k} S^{(n)}(x^k).$$

Noticing that all the negative terms in $-\Sigma(x^{kn}/k)$ are more than balanced by the positive terms in $\Sigma S^{(n)}(x^k)/k$, we conclude that

(24) $$\log(S^{(n)}(x)/x) \geq S^{(n)}(x),$$

or equivalently,

(25) $$\frac{1}{x} \geq \frac{S^{(n)}(x)/x}{\log(S^{(n)}(x)/x)}$$

Since $\eta_{(n)} \geq \eta$ which is known to be at least 1/4, we see that $S^{(n)}(x)$ is bounded on the open interval $(0,\eta_{(n)})$. But $S^{(n)}(x)$ is clearly monotonic, and so the left hand limit at $\eta_{(n)}$ exists, and we let $b_o^{(n)}$ denote this limit. It now follows quickly that $b_o^{(n)} = S^{(n)}(\eta_{(n)})$.

<u>Lemma 6b</u>. The series $S^{(n)}(x)$ satisfies

$$S^{(n)}(\eta_n) = b_o^{(n)} = 1.$$

<u>Outline of Proof</u>. This lemma is proved in the form $T(\eta) = 1$ in Harary and Palmer [8, Section 9.5]. The main steps in the proof, which corresponds in detail to the proof in the reference cited, are:

1. Define the function of two complex variables

$$(26)\quad F(x,y) = (x-x^{n+1})\exp\{y + \sum_{k=2}^{\infty} \frac{1}{k} S^{(n)}(x^k)\} - y.$$

2. Observe that $y = S^{(n)}(x)$ is the unique analytic solution of $F(x,y) = 0$, it has a singularity at $x = \eta_{(n)}$, and furthermore, $F(\eta_{(n)}, b_o^{(n)}) = 0$.

3. Calculate the partial derivative in (26) with respect to y to find that

$$(27)\quad \left(\frac{\partial F}{\partial Y}\right)_{(\eta_{(n)}, b_o^{(n)})} = b_o^{(n)} - 1.$$

4. Conclude that this partial must be zero since $y = S^{(n)}(x)$ has a singularity at $x = \eta_{(n)}$. Thus $b_o^{(n)} = 1$ as required.

We have found that

$$(28)\quad S^{(n)}(\eta_{(n)}) = T(\eta) = 1.$$

Since both series have nonnegative coefficients, and since the coefficients $S_p^{(n)} = T_p$ for $p \leq n + 1$ and $S_p^{(n)} < T_p$ for $p \geq n + 2$, we conclude that $\eta_{(n)}$ is strictly greater than η, completing the proof of the theorem.

<u>Theorem 7</u>. For any rooted tree L of weight n, almost every tree has L as a limb.

<u>Proof</u>. By Theorem 6, we see that the series for trees not having limb L converges on a larger disk than does the series for all trees. Consequently, the coefficients t_p have a faster rate of growth than the coefficients $r_p^{(n)}$. We conclude that

$$\lim_{p\to\infty} \frac{r_p^{(n)}}{t_p} = 0. \tag{29}$$

Aside from being the tool needed to obtain our main result, this theorem is interesting in its own right. Let us apply the theorem to small limbs. Every nontrivial tree has an endline, which is a limb of weight 1, so the first nontrivial application occurs when the weight is 2. Then the two possible limbs as shown in Figure 6 provide two nice asymptotic results.

<u>Figure 6</u>. <u>The two limbs of weight 2</u>.

The next two statements, which follow easily from Theorem 7, are already implicit in Harary and Prins [9] and explicit in Harary and Palmer [8, Chap. 9].

Corollary 7a. Almost every tree has a nontrivial automorphism.

Corollary 7b. Almost every tree is homeomorphically reducible.

Proof. Since almost every tree has L_1 as a limb, interchanging the two endpoints of this limb provides a nontrivial automorphism, proving the first corollary. Similarly, almost every tree has a limb L_2 which provides a point of degree 2. Thus, almost every tree is homeomorphically reducible.

Although these two corollaries have been proved independently [8] (by enumerating trees with identity group and homeomorphically irreducible trees), it is pleasant to see them verified so readily by applying Theorem 7 to the two possible limbs of weight 2. The exact numbers of rooted trees, ordinary trees, identity trees, and homeomorphically irreducible trees has been computed for $p \leq 39$. We find that for $p = 39$, roughly one tree in a thousand has the identity group while only one in a hundred thousand is homeomorphically irreducible. In fact, it has been verified that the expected ratio of these two species of trees approaches zero.

5. Conclusion.

We can now easily prove our main result.

Theorem 8. Almost every tree has a cospectral mate.

Proof. By Theorem 7, almost every tree has the rooted tree R with 8 lines of Figure 2 as a limb. Moreover, according to Theorem 3, replacing one limb R by one limb S leaves the characteristic polynomial unchanged. Furthermore, the new tree formed by this replacement is distinct from the original because it has one fewer limb isomorphic to R than the original did.

If we define $R_i^{(k)}$ for $0 \leq i \leq k$ to be the limb formed by identifying the roots from i copies of S and k-i copies of R, then the k+1 limbs $R_i^{(k)}$ may be substituted for one another as in Theorem 3, i.e.,

$$\phi(T \cdot R_i^{(k)}) = \phi(T \cdot R_j^{(k)}) \quad \text{for } 0 \leq i < j \leq k. \tag{30}$$

We may use this family of limbs to prove a somewhat more general result.

Corollary 8. For each positive integer k, almost every tree has k cospectral mates.

For trees on a small number of points, the great majority are uniquely determined by their spectra, yet Theorem 8 tells us that, asymptotically, this is the exception. How many points p are needed in order to be sure that less than half of the trees with p points are uniquely determined by their spectra? Our method of proof suggests that we estimate the value of p which makes $S_p^{(8)}/T_p \cong 1/2$. Careful estimates of $S_p^{(8)}$ and T_p tell us that their ratio is asymptotic to $(\eta/\eta_{(8)})^p$. These parameters have been estimated by computer, and the result is

$$(31) \qquad \left(\frac{\eta}{\eta_{(8)}}\right)^{4919} \cong \left(\frac{.338\ 321\ 856\ 899\ 21}{.338\ 369\ 674\ 024\ 30}\right)^{4919} \cong .5.$$

Thus, we must have 4919 points to be sure that no more than half the trees are uniquely determined by their spectra. But our method of proof finds only those cospectral pairs which result from replacing a limb R by limb S. Certainly there are many other cospectral pairs not discovered in this way. Thus, the author suspects that really only about 100 points are actually needed to attain the fraction of 1/2. Still, that is such a tremendous size that it is not surprising that earlier investigators had conjectured erroneously that almost all trees are uniquely determined by their spectra.

Does Theorem 8 offer significant evidence on the conjecture for general graphs?

Conjecture. Almost every graph has a cospectral mate.

In general, the asymptotic structure of trees is vastly different from the asymptotic structure of graphs. For example, it is well known see [8 , Section 9.4] that almost all graphs have just the identity automorphism group. Yet, we have just shown that almost all trees do have a nontrivial automorphism. Similarly, it is easy to show that almost all graphs are homeomorphically irreducible, just the opposite of what we found for trees! Thus, we must conclude that the asymptotic result for trees offers no guidance for deciding the spectral uniqueness question for graphs.

Acknowledgement.

The author is grateful to Professor F. Harary for his many helpful suggestions.

REFERENCES

[1] G. A. Baker, Jr., Drum shapes and isospectral graphs, J. Math. Phys. 7 (1966), 2238-2242.

[2] A. T. Balaban and F. Harary, The characteristic polynomial does not uniquely determine the topology of a molecule, J. Chem. Doc. 11 (1971), 258-259.

[3] L. Collatz and U. Sinogowitz, Spectra of finite graphs, Abh. Math. Sem. Univ. Hamburg 21 (1957), 64-77.

[4] M. Fisher, On hearing the shape of a drum, J. Combinatorial Theory 1 (1966), 105-125.

[5] F. Harary, The determinant of the adjacency matrix of a graph, SIAM Rev. 4 (1962), 202-210.

[6] F. Harary, Graph Theory, Addison-Wesley, Reading, 1969.

[7] F. Harary, C. King, A. Mowshowitz, and R. C. Read, Cospectral graphs and digraphs, Bull. London Math. Soc. 3 (1971), 321-328.

[8] F. Harary and E. M. Palmer, Graphical Enumeration, Academic Press, New York, 1973, to appear.

[9] F. Harary and G. Prins, The number of homeomorphically irreducible trees, and other species, Acta Math. 101 (1959), 141-162.

[10] A. J. Hoffman, On the polynomial of a graph, Amer. Math. Monthly, 70 (1963), 30-36.

[11] R. Otter, The number of trees, Ann. of Math. 49 (1948), 583-599.

[12] G. Pólya, Kombinatorische Anzahlbestimmungen für Gruppen, Graphen und chemische Verbindungen, Acta Math. 68 (1937), 145-254.

[13] H. Sachs, Beziehungen zwischen den in einem Graphen enthaltenen Kreisen und seinem charakteristischen Polynom, Publ. Math. Debrecen 11 (1964), 119-134.

[14] S. S. Shrikhande, On a characterization of the triangular association scheme, Ann. Math. Statist. 30 (1959), 39-47.

[15] L. Spialter, The atom connectivity matrix and its characteristic polynomial, J. Chem. Documentation 4 (1964), 261-274.

WHAT IS A MAP?

William T. Tutte*

Maps are usually presented as cellular dissections of topologically defined surfaces. But some combinatorialists, holding that maps are combinatorial in nature, have suggested purely combinatorial axioms for map theory, so that that branch of combinatorics can be developed without appeals to point-set topology. Thus A. Lehman (unpublished) views a map as a permutation acting on the set of directed edges of a graph, and this point of view is adopted in the important enumerative thesis of Walsh [6]. Similar treatments are given by Cori [1], Jacques [2,3] and T. Lenormand and N. Biggs (unpublished).

In the present paper, I investigate the possibility of defining a map as a permutation acting on a set of doubly directed edges.

Let us start with a non-null finite set E of elements called "edges". We suppose each edge to have a set of four associated "oriented forms". No

*This work was supported by a grant from the National Research Council of Canada.

two edges are to have an oriented form in common. The oriented forms of an edge A are to be grouped in pairs in two different ways: in the one case, as two "θ-pairs" and in the other as two "ϕ-pairs". We denote by W the set of all the oriented forms of the edges. Its members are "oriented edges".

On W, we define two basic permutations, θ and ϕ. The first, θ, interchanges the two members of each θ-pair, and the second, ϕ, interchanges the two members of each ϕ-pair. Thus if X is one oriented form of an edge A, then the other three oriented forms of A can be written as θX, ϕX and $\theta \phi X$. We note the identities

$$(1) \qquad \theta^2 = \phi^2 = I, \qquad \theta\phi = \phi\theta,$$

where I is the identity permutation on W.

In a diagram, A is usually represented as an edge of a graph. An oriented form X of A is indicated by two arrows, one drawn along A and the other across it. Then ϕ can be interpreted as the operation of reversing the longitudinal arrow, and θ as the operation of reversing the cross arrow (or vice versa).

We now define a <u>$(\theta - \phi)$-map</u> on W as a permutation M on W satisfying the following conditions:

I. $M\theta = \theta M^{-1}$.

II. If $X \in W$ and k is an integer, then $M^k X \neq \theta X$.

III. The group $[M,\theta,\phi]$ of permutations of W generated by M,θ and ϕ is transitive on W.

This definition is suggested by the following treatment of a convex polyhedron. Let each edge be assigned its longitudinal and cross arrows. Let M be the operation transforming an oriented edge X as follows. It rotates X about the negative end u of the longitudinal arrow, in the direction indicated by the cross arrow, into the position of the next edge incident with u. Thus if Figure 1 represents part of the polyhedron, we have for example

$$MX = \phi Z, \quad M\phi Z = Y, \quad MY = X,$$

$$M\theta X = \theta Y, \quad M\phi X = \theta V, \quad M\theta\phi X = T.$$

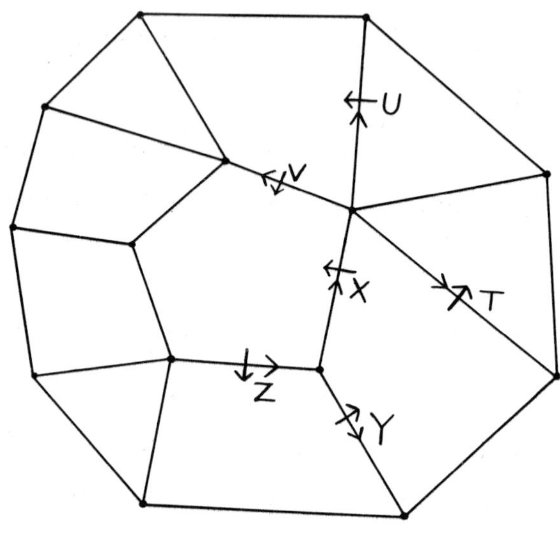

Figure 1

The three conditions are easily verified in this example; the third is a consequence of the connection of the polyhedron.

In a more general topological example, the map is defined by a connected graph G drawn in a topological surface S so as to dissect the remainder of S into a finite number of simply connected domains, homeomorphic to the open disc. The operation M can be defined as in the case of the polyhedron, in terms of longitudinal and cross arrows. There is a slight complication: the graph may have loop. In this case, it is best to think of a loop A on a vertex u as divided into two halves by u and some other point. If X is an oriented form of A then M rotates that half of A in which the longitudinal arrow begins, in the direction of the cross arrow, into the position of the "next edge" at u (which may be the other half of A). The three conditions can still be verified. When X is an oriented form of a loop A, there may be a positive integer k such that $M^k X$ is another oriented form of A, but if so it corresponds to the other half of the loop and so cannot be θX. Figure 2 shows a very simple example of a topological map. Here S is a torus and G has only two edges, both of which are loops.

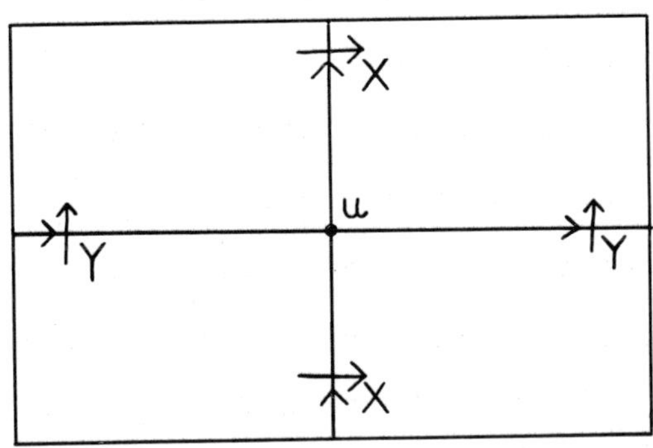

Figure 2

We note that here $MX = \theta Y$, $M\theta Y = \theta\phi X$, $M\theta\phi X = \phi Y$ and $M\phi Y = X$.

On the projective plane we can construct a map whose graph consists of a single vertex and a single loop. The loop is drawn as a non-bounding circuit. It is found that the permutations M and ϕ are identical in this case.

Diagrams such as Figure 2 are like the sketches of curves sometimes found in works on analysis. They are presented as aids to thought, but they are not to be used as steps in proofs. Theorems about maps are to be deduced from the three conditions, and not by the inspection of diagrams.

If $X \in W$ we can construct the "orbital sequence" $(X, MX, M^2X, \ldots, M^{k-1}X)$, where k is the "order" of X, that is, the least positive integer such that $M^kX = X$. The k members of the orbital sequence constitute the "orbit" Orb(M,X) of X in M. This is also the orbit of each member of the orbital sequence of X. The two orbits Orb(M,X) and Orb(M,θX) are distinct and therefore disjoint, by Condition II. The orbital sequence of θX is $(\theta X, \theta M^{k-1}X, \theta M^{k-2}X, \ldots, \theta M^2X, \theta MX)$, by Condition I. Evidently, the permutation θ transforms each of

Orb(M,X) and Orb(M,θX) into the other. We say that these two orbits are <u>opposite</u>.

In our diagrams, an orbit is the set of oriented edges obtained by rotating one oriented edge about a vertex by iteration of the operation M. The opposite orbit corresponds to the same vertex but to the other direction of rotation. This observation motivates the following combinatorial definition.

A vertex of a (θ - ϕ)-map M is a pair of <u>opposite</u> orbits.

The use of the terms "edge" and "vertex" suggests the existence of a corresponding graph. This graph, denoted by G(M) is constructed according to the following rule: the ends of an edge A are those vertices whose constituent orbits include oriented forms of A. Thus if X is an oriented form of A, then one end of A is the unordered pair Orb(M,X), Orb(M,θX), and the other end is Orb(M,θX), Orb(M,θϕX). These two pairs may be one; in that case A is a loop of G(M). The graph G(M) is the one usually shown in diagrams of maps. From Condition III we can deduce that G(M) is connected.

We now come to what I sometimes think of as the Fundamental Theorem of Map Theory.

<u>Theorem</u>. If M is a $(\theta - \phi)$-map on W, then $M\theta\phi$ is a $(\phi - \theta)$-map on W.

<u>Proof</u>. Write $M^* = M\theta\phi$. We have

$$M^*\phi = M\theta\phi\phi = M\theta = \theta M^{-1}, \text{ by Condition I.}$$

Hence

$$M^*\phi = \theta(M^*\theta\phi)^{-1} = \theta\phi\theta(M^*)^{-1} = \phi(M^*)^{-1}.$$

Thus M^* satisfies Condition I for a $(\phi - \theta)$-map.

Suppose M^* fails to satisfy Condition II for a $(\phi - \theta)$-map. Then

$$(M^*)^k X = \phi X$$

for some oriented edge X and some integer k. Since we can add to k any multiple of the order of X in the permutation M^* without affecting the validity of this equation, we may assume $k > 0$. Subject to this condition, choose X and k so that k has the least possible value.

Assume first that $k = 1$. Then

$$M\theta\phi X = \phi X = \theta(\theta\phi X).$$

This is impossible, by Condition II for M.

Now consider the remaining case $k \geq 2$. Here

$$(M^*)^{k-1}X = (M^*)^{-1}(M^*)^kX = (M^*)^{-1}\phi X$$
$$= \phi M^*X,$$

by the result already proved. But then

$$(M^*)^{k-2}(M^*X) = \phi(M^*X).$$

Since this implies that $k > 2$, it is contrary to the definition of k and X.

We conclude that M^* satisfies Condition II for a $(\phi - \theta)$-map.

The group $[M,\theta,\phi]$ is identical with the group $[M^*,\theta,\phi] = [M\theta\phi,\theta,\phi]$ of permutations generated by M^*, θ and ϕ. Hence, by Condition III for M, the permutation M^* satisfies Condition III for a $(\phi - \theta)$-map.

We call M^* the <u>dual map</u> of M. The dual of the $(\phi - \theta)$-map M^* is then the $(\theta - \phi)$-map $M^*\phi\theta$. But this is $M\theta\phi\phi\theta$, that is M. We thus have the identity

(2) $$(M^*)^* = M.$$

The map M^* also has its vertices and its graph $G(M^*)$. We define the "faces" of M as the vertices of M^*. Correspondingly the vertices of M are the faces of M^*.

The faces of M are associated with certain closed paths in G(M). A directed edge, or "dart" is indicated in a diagram by an arrow in an edge of the figure. We make the combinatorial definitions as follows. In a graph G, let A be an edge with ends u and v, possibly coincident. We assert that there are two opposite darts on A. One has "tail" u and "head" v, and the other has tail v and head u. A "path" in G is a finite sequence of one or more darts in G such that the head of each dart, except for the last one, is the tail of its successor. The path is "closed" if the head of its last dart is the tail of its first.

A dart on an edge A is conveniently represented by an unordered pair of oriented forms of A such as $\{X,\theta X\}$. (In a diagram, X and θX have the same longitudinal arrow). Inspection of a diagram suggests the following definition. The tail of $\{X,\theta X\}$ is the vertex $\{Orb(M,X), Orb(M,\theta X)\}$, and the head is the vertex $\{Orb(M,\phi X), Orb(M,\theta\phi X)\}$.

Let f be a face of M, and let X be a member of one of its constituent orbits in M*. Let the orbital sequence of X in M* be

$$X, M^*X, (M^*)^2X, \ldots, (M^*)^{k-1}X.$$

With its member $(M^*)^jX$, we associate the dart $D_j = \{(M^*)^jX, \theta(M^*)^jX\}$ of $G(M)$. Let us denote the tail of this dart by t_j and its head by h_j.

One constituent orbit of t_j is

$$\mathrm{Orb}(M,(M^*)^jX),$$

and one constituent orbit of h_j is

$$\begin{aligned}\mathrm{Orb}(M,\theta\phi(M^*)^jX) \\ &= \mathrm{Orb}(M,M\theta\phi(M^*)^jX) \\ &= \mathrm{Orb}(M,(M^*)^{j+1}X).\end{aligned}$$

Thus the head of D_j is the tail of D_{j+1} if $j < k - 1$, and the head of D_{k-1} is the tail of D_0. Accordingly the sequence

$$P = (D_0,D_1,D_2,\ldots,D_{k-1})$$

is a closed path in $G(M)$.

We call P a "bounding path" of the face f. In the above reasoning, X can be replaced by any other member of $\mathrm{Orb}(M^*,X)$ or of the opposite orbit $\mathrm{Orb}(M^*,\phi X)$ in the $(\phi - \theta)$-map M^*. We thus obtain other bounding paths of f. It is found that all the bounding paths of f are the cyclic permutations of P and its "inverse path" P^{-1}. The inverse path is obtained from P by replacing each dart by its opposite and then reversing the order of the terms.

Perhaps I have now taken the theory almost far enough. In the theory of planar maps, I am prepared to carry on combinatorially when I am given the bounding paths of the faces [5]. However, it is necessary to give some combinatorial recognition to the various "surfaces" associated with maps, and I conclude with a very brief account of this problem.

A $(\theta - \phi)$-map can be described by means of an "orbit-list", a table in which each row is an orbital sequence and there is exactly one row for each orbit. By Condition I, the orbital sequence of θX can be written down when that of X is given. In practice therefore it is only necessary to list one orbital sequence for each pair of opposite orbits. We thus obtain a "reduced orbit-list" of M.

A further simplification can be made by deleting the symbol ϕ throughout the reduced orbit-list. We thus obtain a "scheme" of the map M. In a scheme, each edge appears exactly twice in oriented form. To reconstruct a reduced orbit-list of M, we have to apply the operator ϕ to one but not both of these occurrences. Which one does not matter; no harm is done by interchanging the symbols

X and θX for the two members of a ϕ-pair, provided that the symbols θX and $\theta\phi X$ are also interchanged.

This scheme is identical with the scheme used in the classical topological theory of combinatorial manifolds, except that there it is regarded as describing M* rather than M. (I am thinking in particular of the treatment in [4]). Certain topological operations, subdividing edges and faces of M*, are interpreted as combinatorial operations on the scheme. It is found that each of these operations transforms M* into another map without altering two important numbers called the Euler and orientability characteristics.

The Euler characteristic of a map M is obtained by subtracting the number of (unoriented) edges from the sum of the number of vertices and the number of faces. We denote it by N(M). Clearly $N(M) = N(M^*)$. A $(\theta - \phi)$-map M is said to be non-orientable if the group $[M, \theta\phi]$ of permutations generated by M and $\theta\phi$ is transitive on W, and orientable otherwise. Since $[M, \theta\phi]$ is identical with $[M\theta\phi, \theta\phi]$ it is clear that M and M* are either both orientable or both non-orientable. The orientability characteristic $\chi(M)$ of a map M takes the value 0 or 1 according as M is orientable or non-orientable.

Once subdivision operations are interpreted combinatorially in terms of a scheme, there is no difficulty in representing them as simple operations on the permutation M. Using them and their inverse operations, a scheme is transformed into one of a family of "normal forms". No two normal forms represent maps with both the same Euler characteristic and the same orientability characteristic. The conclusion is that the integer N(M) may be positive, negative or zero, but cannot exceed 2. There are maps corresponding to each value of N(M) from 2 downwards. If N(M) is odd, only non-orientable maps occur, but if N(M) is even, both orientable and non-orientable maps are found.

A combinatorial surface can be defined as a family of maps -- all maps with a specified value for each of N(M) and χ(M). Any two maps of such a surface can be transformed into one another by subdivision operations and their inverses. The sphere is the surface for which N(M) = 2, and the projective plane is the one for which N(M) = 1. On the "torus" we have N(M) = 0 and χ(M) = 0, and on the

"Klein bottle" we have $N(M) = 0$ and $\chi(M) = 1$. A "planar map" is a map on the sphere, and a "planar graph" is a graph isomorphic with the graph $G(M)$ of some planar map M.

The combinatorial foundation is now complete and we can continue to develop a rigorous theory of planar or other maps without appeals to topology.

Before concluding it may be of interest to give a combinatorial definition of the angles at a vertex v of a $(\theta - \phi)$-map M, or in a face of M.

Let X be a member of a constituent orbit of v. We refer to the ordered pair (X,MX) as an M-pair. In a diagram, we can associate it with a directed angle at v, the angle through which we rotate X in order to bring it into the position of MX. The opposite directed angle would be associated with the M-pair $(\theta MX, \theta X)$. Accordingly, we refer to the unordered pair

$$\{(X,MX),\ (\theta MX,\theta X)\}$$

of M-pairs as an angle of M at v. Its arms are the darts $(X,\theta X)$ and $(MX,\theta MX)$, whose tails are at v.

The number of distinct angles at a vertex v is equal to the valency of v in the graph of M. Thus if v is monovalent, the only angle there is

$\{(X,X),(X,X)\}$. But if v is divalent, we have the two distinct angles

$$\{(X,MX),\ (\theta MX,\theta X)\},$$
$$\{(MX,X),\ (\theta X,\theta MX)\}.$$

Let us now relate the angles of M to those of M*. If (X,MX) is any M-pair, we define its <u>conjugate</u> M*-pair to be $(\theta\phi X, M^*\theta\phi X)$, that is $(\theta\phi X, MX)$. Then the conjugate of $(\theta MX,\theta X)$ is $(\theta\phi\theta MX, M^*\theta\phi\theta MX)$, that is $(\phi MX,\theta X)$. Thus the operation of conjugacy converts the angle

$$A = \{(X,MX),\ (\theta MX,\theta X)\}$$

of M into the angle

$$A^* = \{(\theta\phi X,MX),\ (\phi MX,\theta X)\}$$

of M*. It is easy to verify that $(A^*)^* = A$.

The angles in a face F of M can be defined as the conjugates of the angles at the vertex F of M*.

References

[1] R. Cori, Graphes planaires et systemes de parentheses, Doctoral Thesis, Paris, 1969.

[2] A. Jacques, Sur le genre d'une paire de substitutions, <u>C. R. Acad. Paris</u>, 267 (1968), 625-627.

[3] A. Jacques, Constellations et graphes topologiques, <u>Colloq. Math. Soc. János Bolyai</u> 4, <u>Combinatorial Theory and Its Applications</u>, Balatonfüred (Hungary), 1969.

[4] H. Seifert and W. Threlfall, Lehrbuch der Topologie, Leipzig, Teubner, 1934.

[5] W. T. Tutte, On the enumeration of planar maps, Bull. Amer. Math. Soc., 74 (1968), 64-74.

[6] T.R.S. Walsh, Combinatorial Enumeration of Non-planar Maps, Thesis, University of Toronto, 1971.